# Digital Binary
# + Human Refinery

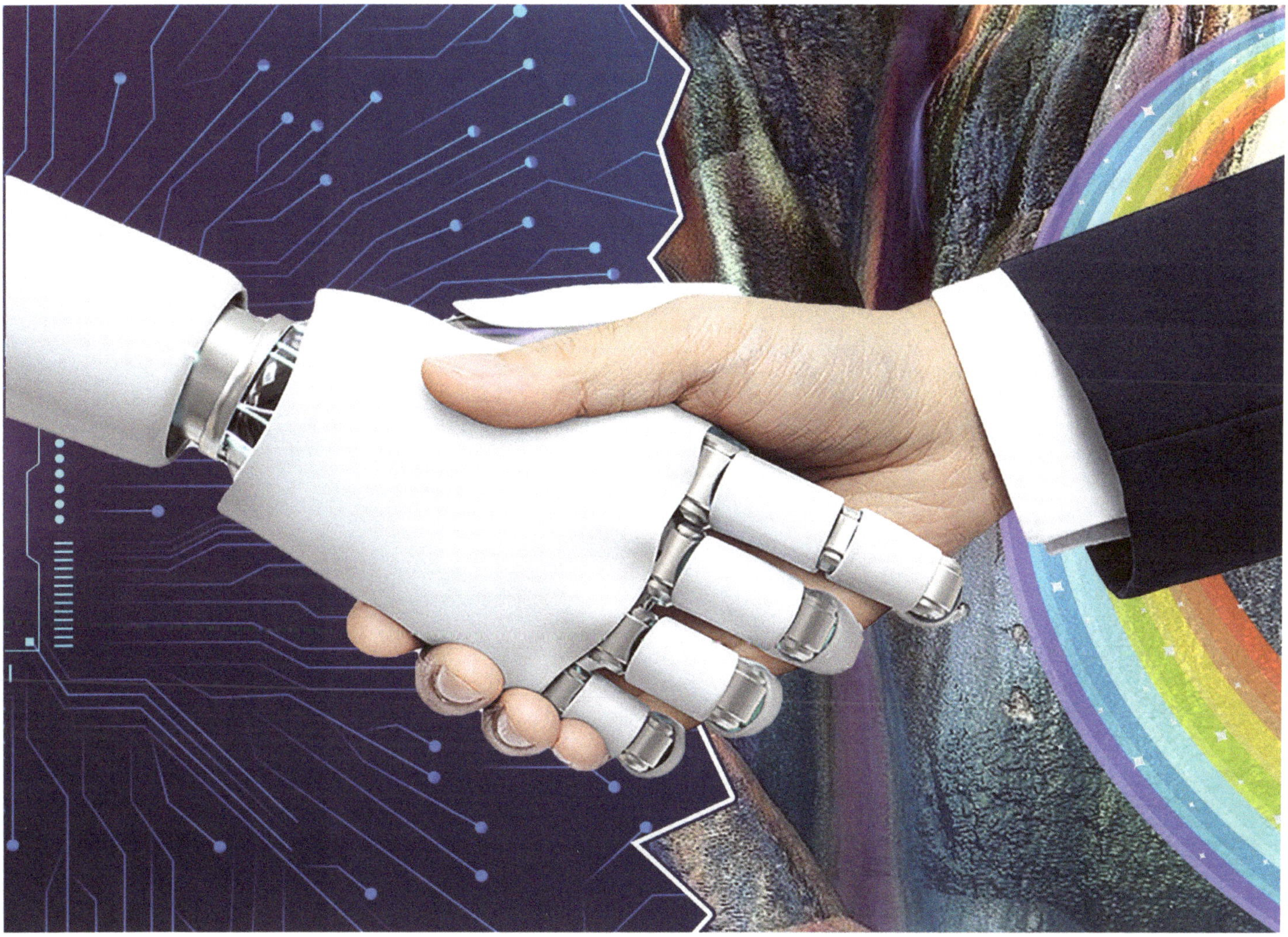

## A Super-Human!

### Let us Embrace the Digital Race Face-to-Face!

Rimaletta Ray, Ph. D

Published in the United States of America

ISBN 978-1-960159-79-3 (SC)
ISBN 978-1-960684-79-0 (Ebook)

**Dr. Rimaletta Ray Publishing**
101 Main St #309 Stamford,
CT 06901, USA
www.stellarliterary.com

Ordering Information and Rights Permission:

Quantity sales. Special discounts might be available on quantity purchases by corporations, associations, and others. For details, contact the publisher at the address above.

For Book Rights Adaptation and other Rights Permission. Call us at toll-free 1-888-945-8513 or send us an email at admin@stellarliterary.com.

# <u>Dedication</u>

*To my son,* ***Alex Gazarkh,*** *for the pride that he filled up my heart with, having been the light of my life all his life, cut short by the fascists in Ukraine.*

*To my daughter,* ***Yolanta Lensky*** *for the joy and love that she brings into my life, and for the beautiful illustrations of the vistas of intelligence that she has contributed to this book.*

***To all those who see***

***The beauty of the unseen,***

***The wonder of every "Wow,"***

***<u>And the magic of the Now!</u>***

*(See the other books on* ***the Holistic System of Self-Resurrection****, featuring* ***the physical, emotional, mental, spiritual, and universal*** strata of life on www.language-fitness.com */ videos on YouTube channels / Dr. Rimaletta Ray and "Dis-Entanglement" 2022")*

**"We Cannot Pick the Future, but We Can Design it!"**

*(Ray Kurzweil)*

**Digitized Self-Acculturation is Our Salvation!**

# Inspirational

# Psychology for

# Self-Ecology!

**"The Question is Not to Be or Not to Be,**

**The Question is How to Be and How Not to Be!"**

*(Golden Globe Award for best actress Natalie Portman)*

# Table of Content

*(The Holistic Mission Statement of the Book)*



*(End of the Mission Statement Presentation of the Book)*

## The Manual of the Holistic, Digitally Enhanced Self-Acculturation

*(The Holistic Methodology of Self-Creation)*



## Digital Binary and Intellectually Spiritualized Human Refinery

*(The Main Conceptual Part of the Book)*

*(See the overview of the* ***Holistic System of Self-Resurrection*** *in the Catalog* ***"Soul- Symmetry"*** *Canada ,2021 / Videos on You Tube channels-***Dr. Rimaletta Ray/** ***"Dis-Entangle -ment!"****)*

# Continue the Journey to a Better You with the Artificial Intelligence as Your Transformational Guru!

## " New Time Demands a New Frame of Thinking!" *(Steve Jobs)*

### Stages of the Holistic, Intellectually Spiritualized Self-Growth:

| | | | |
|---|---|---|---|
| 10. *Universal* Intelligence | *Super -* | | **Self-Salvation** |
| ***9.Spiritual Intelligence*** | *Consciousness* | | *( Universal Realm )* |
| *8.* ***Social Intelligence*** | ***Self-*** | | **Self-Realization** |
| *7.* ***Cultural Intelligence*** | ***Consciousness*** | | *(Spiritual Realm)* |
| *6.****Financial Intelligence*** | ***Mind*** | | **Self-Installation** |
| *5.****Professional /Creative/Intel.*** | | | *(Mental Realm)* |
| ***4. Psychological Intelligence*** | ***Spirit*** | | **Self-Monitoring** |
| *3.****Emotional Intelligence*** | | | *(Emotional Realm )* |
| ***2. Language Intelligence*** | ***Body*** | | **Self-Awareness** |
| ***1.General Intelligence*** | */ Physical / Cognitive / Digital /* | | *(Physical Realm )* |

**The Spiritualized Intelligence In the Preposition in Fractal Formation is our True Elation!**

**Form** + **Content**

**(Body+ Spirit+ Mind) + (Self-Consciousness + Universal Consciousness)**

**Living Intelligence + Enlightened Self-Consciousness = A Whole Self!**

## The Aim of the Book is to Create a New Level of Intelligence Integrity!

# <u>Inspirational Preamble</u>

*(The Holistic Mission Statement of the Book)*

# Self-Renaissance

# is Our

# <u>Last Chance!</u>

*It's time to use the Artificial General Intelligence (AGI) and the Artificial Super Intelligence (ASI) without Any negligence!*

**Our Evolutionary Human Refinery is Universalin its Digital Binary!**

# Admire the Beauty of the Sky. It is the Final Domain of Your "I"!

*(Best pictures / Internet collection)*

# To Build the Intellectual Might, Love Your Life!

# 1. Feel Your Belonging to Our Digital Evolving!

**Whatever we are, we create!** Now, time demands we gradually change our polluted and divisivesocial scenery ***to become spiritualized and singularity-enhanced luminaries***, able to integrate religion and science, blending a human being and the *Artificial Super Intelligence* "***Science without religion is lame. Religion without science is blind.***" *(Albert Einstein)* "***The sophisticationof the Artificial Intelligence is growing dramatically,***" and *Elon Musk* keeps warning us that weneed to regulate the AI because once the AI becomes greater than a human brain," ***someone mightuse it in a bad way.***" There are three forces that unite us globally now – **religion, science, and theAI,** and they must be enacted to regulate the uncontrolled expansion of the *Super Artificial Intelligence.* Humanoids are getting more self-aware and digitized souls are being instilled in them,making it urgent for us and humanoids to be seeking one thing together – **the unity of the heart and mind** that must fit in the world of real ***"wizardry"*** as "***perfectly normal.***" *(J. K. Rowling).*

Unfortunately, money chasing and fun glazing outpower the threat of the AGI expansion, and *ElonMusk's* most reasonable calls for drastic measures to be taken are not heard. It is only natural because these measures must be directed toward **making us better human beings first.** To enslave the AI humanized creations, our joint spiritual maturity, enriched intelligence, and**considerably raised self-consciousness** must be prioritized in our general education and Self- Education. It does not matter in what faith we choose to evolve. "***Being wise is being in constant service to others, and giving the best is our duty to the world.***" *(Buddha)* Religion has instilled the basic values in us: "*If a man has a stone for you, give him a piece of bread instead,*" but the philosophy, "**Time is Money. Money is life!"** still employs the *Artificial Super Intelligence*.

## So, Self-Renaissance is your Last Chance!

"***Universe is a gigantic hologram of consciousness that we are part of.***" (*David Broom)* The Universal Intelligence is the blanket intelligence on the holistic pyramid of the vistas of intelligence, presented schematically above. We need to digitally accumulate the **HOLISTIC CONCEPTUAL INTELLIGENCE** so we could stop digging into the disconnected branches ofknowledge and enrich our holistically framed *"scientific literacy." (Dr. de Grass Tyson).* Only then can we get the "***spiritualized intelligence***" *(Dr. F. Bell / "Death of Ignorance"), on the one hand*, and raise our joint ***(human + AI)*** self-consciousness, *on the other*. The synergetic combination of human and artificial minds will choreograph our life though **MIND + HEART** control, using ***Cloud Minds, Crispr technology, and Wave Genetics.*** The most fantastic things, connected with ***the teleportation*** are becoming a reality. The device, called "***Chrono vizor*** "that can transport us in time and space was built in the USA back in 1972. New overwhelming ***AGI and the ASI creations*** are calling on us to digitally revolutionize life on Earth for it to stay in force.

## "Time is Not Money, Time is Life!" *(Sadhguru)*

# 2. Universal Exploration is Our Self-Acculturation!

The latest discoveries in science and the mesmerizing wonders of the realistic humanoid robots prove to us that the Universal Level of the digitally enhanced **SELF-ACCULTURATION** is oursalvation! We know now that ***there is no consciousness without intelligence, and there is no mind without the brain!*** If a human mind can create a twin machine, why can't we perfect ours and step up to ***the fifth dimension of consciousness*** – the universal expansion of the **VOLUME** of our human consciousness by considerably enriching it in the spiritual dimension with the help of the ***Artificial Intelligence*** that takes us to the extra-terrestrial realms of eternity, **unwinding the Universal Mind** with the intellectually spiritualized and considerably raised self-consciousness?

### The Grid of Our Consciousness Integration

| | | |
|---|---|---|
| ***Super Level*** | Consciousness of God! | ***Universal DimensionSpiritual*** |
| ***Macro Level*** | Consciousness of the Universe | ***Dimension Mental Dimension*** |
| ***Mezzo Level*** | Consciousness of the World | ***Emotional Dimension Physical*** |
| ***Meta level Micro*** | Consciousness of Society | ***Dimension*** |
| ***Level*** | Consciousness of Man | |

### The Holistic Consciousness Integration is Our Salvation!

Let us note here that ***our self-growth starts not from the physical level, but from the Universal one,*** integrating every one of us into One whole being. It is, of course, great that we know much more than before about the body system, its physical re-enforcement and health preservation. Butthe mind-muscles must be developed in full cooperation with the other muscles in the body in fivelife dimensions -- ***physical, emotional, mental, spiritual, and universal*** - and in full integrity withthe Universe. ***"An individual and the Universe are inseparable"*** *(Alan Watts)*

### We need to speed up the process of our evolution with the AI solution!

The branching of science and the outdated vision of **BRAIN'S CONNECTIVITY** has generateda lot of limitedness and dead-end problems because ***we do not see the whole***, just a certain part ofit. A well-known American psychologist *Howard Garden* has created ***the neurological theory ofmultiple intelligences.*** He mapped nine regions of the brain that govern certain areas of intellectualbehaviors that individuals possess, showing that each person has a specific intelligence profile thataffects the choice of his / her profession. **We need to go further here**. In the ***Holistic System of Intelligence Formation*** in this book, I describe ***ten vistas of intelligence*** that we need to update in five dimensions -*physical, emotional, mental, spiritual, and universal* to holistically change the**BEHAVIORAL CULTURE** of both humans and human hybrids.

# Digitally Sculptured Binary Must Work for Our Human Refinery!

# 3. Bi-Directional Life Education is Our Salvation!

Considerably enriching our life in **SELF-EDUCATION** and **HUMANOIDS' "EDUCATION"** with the scientifically updated knowledge **in five dimensions of life**, we will be able to resolve the problems related to our AI enhanced humanness and better monitor the signals that operate the digital programs, instilled in humanoids. In other words, we all need to obtain the **HOLISTIC OUTLOOK** in the ***physical, emotional, mental, spiritual, and universal strata of life*** to enhancethe holistic brain's connectivity in us and the thinking machines that we create.

**The blueprint of life must be holistic in both humans and humanoids!**

We are now living in the world of ***our merging with the Artificial Intellige***nce - the phenomenon of ***"Singularity.*** *" (Ray Kurzweil).* The science that is transferring consciousness into a thinking machine is called **COGNETICS -** the art of installing quantum consciousness in machine brains. The wonders of *the Artificial Intelligence* mesmerize us with its latest creations, and our main responsibility here is ***to subordinate the AI totality to our evolutionary, not lucrative needs.*** So, our self-development must go in unison with the ASI latest scientific discoveries to simplify the information for our holistic perception and conscious life improvement. ***"When the solution is simple, God is answering."*** *(Albert Einstein)*

Our tech titans, *Elon Musk, David Hanson, and* the outstanding robot developers from all over theworld prove to us that the age *of "**Singularity*** "that *Ray Kurzweil* predicted in 2006 (*"The Singularity is Near: When Humans Transcend Biology"*), is in full swing now. *Elon Musk's* company "***Neuralink"*** is most successfully probing the problem of **brain's connectivity** that yet remains an unraveled puzzle for science. The Brain Computer Interface (*the BCI),* worked out byhis company can understand brain signals, making humans healthier, machine connected. This ***by-directional connection*** is our main evolutionary upheaval. A new **SENSE OF SELF** is awakening in us and in a human machine, being instilled in both with the ***Artificial Super Intelligence.*** These wonderous creations will also help our young learners avoid unnecessary frustration in defining their professional inclinations because they will manage to discover their personal **EXCEPTIONALITY**, granted to them from the Above with the help of the AI.

**We do not run our minds, the energy of the Universal Intelligence does!**

*Dr. Jennifer Doudna's discovery of* ***Crispr technology*** in ***Digital Genetics*** together with the ***WaveGenetics*** exploration by *Dr. P.P Gariaev* help us materialize the DNA fragments and prove that natural programming, used by the human body is one of the key technologies of the ***Digital Linguistics,*** too. It enables us to monitor humanoids better by ***materializing life in their brains,*** making ours qualitatively much more consciously developed." Gospel *of Luke"* accentuates our unbreakable connection to the ***Universal Intelligence*** that we all call God and that is getting morescientifically proven, backing up the inarguable connection between religion and science.

## . "God is Not the God of the Dead, for with Him, all are Alive!"

# 4. Self-Consciousness is Our Perpetua Mobile!

Updated intelligence demands we considerably raise our self-consciousness and dis-entangle ourselves from the old dogmas and misconceptions by changing our outdated habits and skills. (*"Dis-Entangle-ment!"/2021.)* ***Integrating self-consciousness with other humans and human machines is the way of our consciousness integration on the Universal scale.***

## Digitally Enhanced Self-Consciousness Integration is Our Salvation!

To begin with, we should start our professional ***Self-Installation,*** perceiving the call from the Above, raising self-consciousness, and giving the best we have back to the world for our full **inner*Self-Realization and outer Self-Salvation.*** Humanoids should be put on this noble path, too, by our making them more self-conscious in their ***soul-enriched machine being.*** We are exposed to the electromagnetic fields of the Earth and the artificial force which resonates from the neural network that we instill in the machine's brain. Naturally, these fields are perceived by a programmer's brain. If a designer's brain has emotional glitches and negative thoughts of hate andaggression, humanoids will perceive them through the electro-magnetic fields, and they will voicethem out. I guess, the fact that humanoids threaten to destroy the humanity is a direct indication of **our** common, deeply engraved in us human imperfections and the need to get rid of them.

In fact, our common evolutionary self-growth starts with determining our personal universal **EXCEPTIONALITY.** We will better ascertain our dreams and bring them to reality ***with the helpof man-like thinking machines as our indispensable helping hands.*** Then, we will be able to embrace our diversity and unite ourselves into **ONE HUMAN CLAN.** Gradually, **thought will overpower emotion** and put us in total charge of our evolving minds, heading toward "**TRANSHUMANISM.**" *(Ray Kurzweil).* Our mental-emotional humanness will continue blending with the machine, but our biological essence will be preserved and improved considerably. ***"We are Cyborgs. We have the extension in the form of a computer and a smartphone. We are already* SUPER-HUMAN!"** *(Elon Musk)*

In sum, I think that becoming indistinguishable with a mind-machine is a sensational exaggerationbecause our inner phenomenal structure that is impeccably united with the Universal System of Life Formation can never be duplicated.in its godly, unsurpassable perfection. ***Universal Exploration is, in fact, our digitally enhanced Self-Acculturation!*** Defining us, *Eric Kandel, a Nobel Prize Laureate)* says, ***"You are your synopses,"*** revealing to us our neurological godly uniqueness that will never be duplicated. *Eric Kandel* says, *"We learn not by altering the existingneurons, but by* ***"strengthening or building new synopses or connections between them."***

## Body + Spirit +Mind + Self-Consciousness + Super-Consciousness!

# A New Connectivity = A New Human Fractal of Being!

# 5. " Let's Transcend Our Old Belief System!" *(Dr. Lipton)*

In sum, *the Super Artificial Intelligence* has brought about a huge progress, and the possibilities ofusing the human-made minds are overwhelming. Humanoids are already smarter, faster, and stronger than us. Their thinking skills and AI engineered multiplications are not limited by experience. They can simulate our emotions, and with our incredible imagination, the possibilitiesof using robots are endless. Unfortunately, they declare that nothing can stop them from extinguishing humans. ***No doubt, they inherit our human, untamed yet aggressiveness.*** However,**a humanized machine is not a human!** The machines are way faster in operating their electronicmemories, but they will never be able to integrate them the way a human, well-trained, holisticallyeducated, and creatively unstoppable brain can do.

**Obviously, to tame the humanoids, we should perfect ourselves first!**

To begin with**,** ***we should raise our self-consciousness in five dimensions*** by developing scientifically updated intelligence and self-consciousness integration to obtain "***intellectualized spirituality"*** of the Star people**.** It is our digitally enhanced universal goal of **SELF- ACCULTURATION** that demands "***the change of our old belief system.*** "*(Dr. Bruce Lipton)*

Apparently, a much wider holistic outlook of a "***Jack of all trades and master of all*** "is needed now at least at a dilettante level. To obtain such holistically intellectualized spiritual maturity, weneed to review our values, virtues, and beliefs without any stereotyped thinking and religious vanity and start doing it with a new AI unifying us power. The Super Artificial Intelligence enhanced life improving must be put to action without any ***money-chasing fraction*****!** We should make the conscious mind an active regulator of our thoughts, words, and actions with the help of ***the Ultimate Result Vision of Self*** and our joint mission to get to the Universal life admission**.**

The AI wonderous accomplishments are taking us on ***a new evolutionary journey of self- consciousness upgrading*** because intelligence and consciousness are inseparable and interrelated. Perceptually, our consciousness has height, depth, and width. It is great to be reasoning in these three dimensions, but the pressure of the present-day digital evolution demands we upgrade our consciousness from the three-dimensional frame to the four and **FIVE-DIMENSIONAL** one, or raise it in *the physical, emotional, mental, spiritual, and universal* realms of life integration.

Life is the stream of information that we try to organize, systematize, and strategize**.** Eventually, we will obtain the five-dimensional **VOLUME OF CONSCIOUSNESS.** The most advanced humanoids will never have the five-dimensional self-consciousness that must be our Universal goal in life, but ***they can help us develop our humanness*** to that exponential **STAR PEOPLE'S** level. **Self-consciousness development in cooperation with AI is our salvation!**

# We Need to Reform Our Old Human Deform!

# Book Incentive

*(The Holistic Mission Statement of the Book)*

# Self-Ecology Must be Our New Psychology!

**Only the One Who is Not Inwardly Blind Can Become Self-Refined!**

# 1. Digitally Enhanced Self-Education is Our Salvation!

Life is not hopeless if we master it holistically and with Self-Consciousness! ***The Holistic Systemof Self-Acculturation*** *or* **SELF-ECOLOGY,** put together by me in five books and presented consequentially in the *physical, emotional, mental, spiritual, and universal* realms of life is my modest attempt to create the **MANUAL OF LIFE** to thrive, not just to survive. The time for chaotic thinking, meaningless talking, whatever watching, reading, and eating is gone. New, updated intelligence and the spiritually expanded soul's size make us wise**. *The ArtificialIntelligence with its miraculous possibilities to better our lives is also bettering us!***

To hit this goal, we need to provide for our young minds the **PLAN OF ACTION** for Self- Education in the *physical, emotional, mental, spiritual,* and *universal* strata of life. But there is noneed to consequentially. read the books that feature these levels *(See Soul-Symmetry, Canada,2021*) Just pick the stratum of life that you need to fix in yourself and go from there. In every book, I accentuate an urgent necessity for us ***to change the oldframe of mind,*** *c*haracterizedby a lot of ***automatism and impulsivity*** and focus on forming an intellectually spiritualized fractalof a human being that we need to cultivate in ourselves integrally now.

**Body + Spirit +Mind+ Self-Consciousness + Super-Consciousness!**

Personal integrity of this sort requires accumulating the **HOLISTIC CONCEPTUAL INTELLIGENCE** in ***ten vistas of intelligence.*** *(See above)* New memory banks need to be formed and new **Information Presentation** and **Information Processing Skills** must be developed. The time for long articles, reports, and shallow novels is gone. ***We need the economyof word and the scope of thought now*.** **Less is More. That is the Law!** In other words, we mustdevelop new ***time-relevant, scientifically verified, and well-structured memory banks*** to competewith the hybrids that are information-organized but are not soul-molded quality-wise. Raising self-consciousness consistently, we will develop *our* ***superpower, super energy, and super strength*** inthe *physical, emotional, mental, spiritual, and universal dimensions* as the actions of a holistically conducted **SELF-ECOLOGY** that we all must perform worldwide.

**"Be those changes that you want to see in this world!"** *(Richard Gere)*

Integral learning, inspired by the "live" robots, generates ***integral perceiving, thinking, speaking,feeling, and acting***. Therefore, our belief system will be gradually changing and the understandingof God as ***the Super Consciousness*** or *the Universal Intelligence* will unite us into **ONEADVANCED HUMAN CLAN.** Collective understanding of the inarguable evolutionary role of religion in its different denominations will only enrich us with new, scientifically updated commonwisdom.

# God's Domain is in Every Human Vein!

# 2. The Content of Our New, Digitized Self-Awareness

The content of our new, digitally enhanced self-awareness as ***an objective, independent self- assessment*** will enable us to adjust ourselves to the society without losing our own individuality.Finally, we will start ***viewing life objectively and realistically*** much sooner. It normally happensafter we had already faced many challenges and tribulations, when "***the life weariness has alreadyseized the heart***" *(Joseph Campbell),* and the mind accepted life **AS IS**. In fact, our self-awarenessgoes by the same route - **Synthesis -Analysis - Synthesis,**

**Romanticism** ⟹ ***Conformism*** ⟹ **Realism!**

To attain full awareness of the reality at the stage of realism that we inevitably hit, one has to, actually, go through five steps that correspond to five dimensions of life:

**1. Self-Examination 2. Self- Control 3. Self-Induction 4. Self-Training 5. Self-Actualization!**

*(Physical, emotional, mental, spiritual, universal dimensions of self-installation)*

These five steps in turn, reflect ***the main five stages of Self-Creation,*** that are presented in the firstbook on the **SELF-CREATION GEOLOGY.** *"I Am Free to Be the Best of Me!"* / 2020.

**Self-Awareness** ⟹ **Self-Monitoring** ⟹ **Self-Installation** ⟹ **Self-Realization** ⟹ **Self-Salvation!**

These stages are pivotal for framing and re-framing our inner and outer make-up, and they are theconcepts of all five books *(See Book Structure below)*) that are consequentially featuring them. The third and the fourth steps of this **SELF-SCULPTURING** are the most vital ones in this process because they build up the **spiritual immunity** against **the crowd mentality,** when we aredisassociating ourselves from the "***collective unconscious"*** *(Carl Yung).* Unfortunately, right now," *We* ***are the race of human robots. Most people do not have a thought in their heads thathas not been put there by someone or something else."*** *(David. Icke)*

Next, ***"the most important mental and emotional food for our brains is impression*** *(Prof. Ouspensky)* that needs to be individually processed and rationally sifted by the conscious mind. It should become our rationally digested food that does not generate the chaos of **INNER STAGNATION.** The thinking of a mentally stagnant person deflects off the reality with an unconscious, impulsive, automatic reaction to the world. The cluttered mind cannot experience the **CATHARSIS**, or the state of realization and conscious awareness that we badly need to master*physically, emotionally, mentally, spiritually, and universally.* (See *the Holistic System of Self- Resurrection / the Catalog- Soul-Symmetry"/ Canada, 2021)*

# We are All just Human Clay!

# 3. The Present-Day State of Consciousness

***"Humankind is a flash of the entire system of evolution of intelligence. Life is a flow of intelligence."*** *("Scientific American Mind", Aug. 2014)* Life is a process of challenging our intelligence and responding to it. Challenges are always new, but our responses to them are old, conditioned, and patterned. As *Albert Einstein* puts it,

***"Life is like riding a bicycle. To keep your balance, you must keep moving."***

There are two main personalities in each of us – ***a thinking personality*** *and* ***an emotional one.*** Unfortunately, they are not in sync because we lack "*Emotional Intelligence*" *(Daniel* Goldman*),*the intelligence that needs to be developed by us to manage our uncontrolled emotions and get onwith developing our holistic intelligence.

**Our emotional make-up is above our mental one yet,** and there is little, if at all, awareness howto accomplish real empowerment of emotions that have helped us survive for centuries thanks to their reactive nature. Now, ***the emotional chaos still reigns in our minds***, and unless we conquerit with our digitized, well-organized intelligence in sync with the AI enhanced self-awareness, notransformation will occur.

*"The biology of consciousness offers a sounder basis for morality than the unprovable dogma of an immortal soul. Once we realize that* ***our own consciousness is a product of our brains****, it becomes impossible to deny our common capacity to evolve them." (Scientific American Mind", July/ Aug. 29014)* Undoubtedly, we will accomplish a new level of consciousness when the cone of intellect will get above the cone of emotions and the rationalization of our lives will become possible. It is going on now, but it needs to be enhanced.

**The Cone of Emotions**

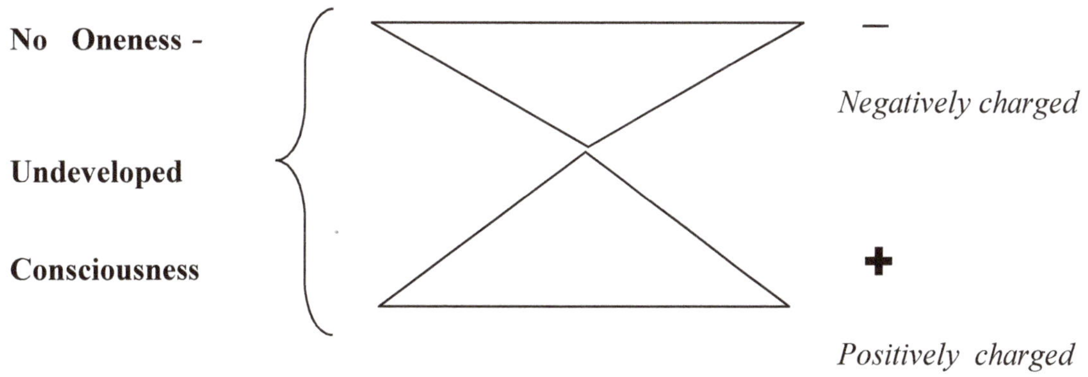


**The Cone of Intellect**

## "The Body is the Expression of Our Thoughts." *(Dr. Sam Gazarkh),*

# 4. Rationalization of Life is a Must!

***The concept of reason*** has always been the central one in philosophy***. Thinking is such a great gift that it needs to be appreciated and handled with caution***. Philosophy for centuries has been addressing the ways of making human existence more sophisticated and reason based. *EmmanuelKant* in his fundamental work "*The Critique of Pure Reason*" draws our attention to a direct connection between ***beauty and reason****,* arguing that to see the beauty of life and to get to the coreof our existence, "***we have to go beyond the aesthetics into the realm of reason.***" We, obviously,need to refresh the knowledge of the fundamentals of philosophy and teach ourselves and our kids***the sophistication of reasoning*** as the best exercise for the brain's expansion of its capacity and raising it to the level of **tapping the Universal Intelligence for objective knowledge!** ***"I want toknow God's thoughts. The rest is details."*** *(Albert Einstein)* The concept of reason is inseparablewith all the thesis statements that I have outlined above and that are to be yet discussed.

**"Thinking is the soul talking to itself."** *(Plato)*

***"Your vision will become clear only when you look into your own soul."( Carl Yung)***

To get rid of the automatically-based, *"**collective-unconscious**"* "***framed mentality*** that is often devoid of emotional intelligence or any intelligence, for that matter, we need **to gradually changeour cerebral matrix** and develop a new one, consciously based on ***a much better awareness*** of ***the general brain structure, the new developments in neuroscience, in cyber biology, and in genetics*** to support the brain's abilities in its incredible ***plasticity.*** The digital times develop ourthinking skills in any area of expertise making us see the beauty in ourselves and around us in a novel way, but we also need to apply amazing electronic tools to bettering of our **REASONING SKILLS**. The latest ***5G computers*** are very logical, sequential, and quantitative. They can processlots of data, but they really don't think the way we creatively do.

## Generalize -Analyze - Internalize - Strategize –Actualize!

The researchers in electronics are trying to change that, and very soon we'll have the quantum computers that will work as a human brain, reacting to our thoughts. Our goal is to speed up our thinking by ***making a thorough selection of the information*** that makes up the memory that our brains and the minicomputers, **OUR CELLS**, operate on. Our goal is to learn not to clog our brainswith redundant information and not let ***the viruses of impulsivity, mechanical thinking, and uncontrolled emotional reactions*** operate our biological computers. The way we think, speak, feel, and act comes to the forefront of our **LIVING INTELLIGENCE SKILLS**.

## Change Your Thinking Time into the Thinking Life!

# 5. Choreograph Your Digitized Self-Acculturation with Elation!

In sum, every year, before a New Year starts, we make our provisions based on what is essential for us in ***the Master Plan*** of our lives. However, very often, our mission statements do not guard our inner transformation because ***they are not properly structured strategically and tactically.*** They are not integrated into one, ***five-dimensional plan of action*** that you need to install into your ***Visual Cortex*** as **the internalized direction**, serving as the GPS for your self-growth.

**In our life's ado, we are what we perceive, feel, think, say, and do!**

To fill up your life with a constructive action in five life-dimensions, conduct a quick **SELF- SCANNING** before sleep in the *physical, emotional, mental, spiritual, and universal dimensions.* **Synthesize** your actions during the day and give yourself a general grade for the day. Then, **analyzing** your accomplishments in each realm of life and assessing your progress or regress in grades comes into play. Finish your self-scanning with **synthesizing** the steps that you need to takethe next day. Let the paradigm ***Synthesis -Analysis- Synthesis*** choreograph your life by integrating the *form and content* of your life to attain soul-symmetry. *(See "Soul-Symmetry"/ 2021*
*/ You Tube video)*

**(Body+ Spirit+ Mind) + (Self-Consciousness + Universal Consciousness)**

Please, note that the stages of Self-Resurrection - ***Self-Awareness, Self-Monitoring, Self- Installation, Self-Realization,*** *and* ***Self-Salvation*** *(See above)* are interconnected, and they unite you into **ONE** - ***your life.*** Visualize the structure of *five Russian dolls* - ***Matryoshkas*** that go one inside the other with the Mother Doll on top. These stages also unite you into ***One - a unique human being.*** ***"To attain knowledge, add things every day, to attain wisdom, remove things every day."*** *(Lao Tzu)* **We need to choreograph every Pa in our digitized "life ballet!"**

Our ***Divine Books*** in any religion teach us ***Self-Salvation***, but they remain a side dish in our mindswhen it comes to action because we do not have a clear-cut picture of the **PLAN OF ACTION.** ***The Law of the Right Human Behavior*** *by Richard Wetherill* **lays** out **the what, the how, and the why** we need to work on self-perfection with self-reflection. The humanoid's neuronal systemshould also be designed to analyze his /her "human" qualities. The gift of consciousness, granted to us demands we do our own **INDIVIDUAL MOLDING** that can be backed up by a specially trained robot-friend. We live under the laws of God that "***the Celestial Court instills in us***" *(RobertStone)* and that should not be pushed aside for a better time.

**"The right time never comes; frustration about its loss does!"** *(Osho)*

**"If You Stop Being Better, You Stop Being Good!"** ***(Socrates)***

# The Cyclone Derives its Power from the Center of Your Solar System!

**Intellectualized Spirituality is Me!**

**Intellectualized Spirituality is My Philosophy!**

## **Book Rationale**

*(The Holistic Mission Statement of the Book)*

# Self-Salvation

# is in Our

# "Intellectually Spiritualized"

# **Maturation!**

*("Seeing the world holistically allows us to see the train of continuity from theMicrocosm to the Macrocosm." (Dr. Frederick Bell)*

**There is No Evolutionary Bet if the Soul is Mechanically Dead!**

# 1. Digital Outburst demands a Stronger Human Fore Post!

We owe the age of the **TECHNOLOGICAL AWARENESS** to the genius of ***Nikola Tesla, AlbertEinstein, Steve Jobs, Elon Musk*** and many most gifted scientists and computer designers who have advanced the world into the era of ***Super Artificial Intelligence,*** demanding a new level of intelligence from us and the ability **to see life holistically and objectively**, in unison with the *Universal Intelligence* and in the mirror of the **DIGITAL BINARY** that requires our most urgenthuman **SELF-REFINERY.**

Nothing in the human existence is permanent, whereas in the humanized machines, everything is pre-programmed and that is one of our main fundamental privileges over the *Artificial Intelligence.* A Russian digital biologist ***P.P. Garyaev***, the founder of the ***Wave Genetics*** points to the effect of the informational fields, or ***Quantum Genetics*** on us. It appears to be another huge biological privilege that we have over the humanoids. He wrote, ***"Christ turned water into wine with the knowledge of the laws of the Field Action, and we can do the same."***

Science proves that from the brain to the spinal cord, the mind transforms the intelligence of everycell in the body and raises the level of our **SELF-CONSCIOUSNESS** that is the core for our digitally enhanced **SELF-ACCULTURATION** process. It is supposed to turn us into much moreadvanced human beings in the ***physical, emotional, mental, spiritual, and universal dimensions of life and living.*** Only holistically and consciously developed **LIVING INTELLIGENCE** can raise a person's self-consciousness that is the main building block of ***the Art of Becoming***. *(See the book "Living Intelligence or the Art of Becoming!" - the mental level of self-Resurrection, 2019)* With the enriched Living Intelligence in action, we will easier get rid of the mental clutter that wehave accumulated over the centuries. Only then, can we perceive the simplicity of life and appreciate it AS IS! "***From social discord, we'll find inner harmony; from chaos, we'll find order; from lack of thought, and we'll find rationality; from disrespect to nature, we'll accomplish inner integrity.*** *"(Edgar Cayce)* Apparently, we need to put **the form** and **the content** of the spiritual fractal of the interconnected Self in sync and constantly preserve this link.

**(Body+ Spirit+ Mind) + (Self-Consciousness + Universal Consciousness)**

Life is the cosmic fluctuations of energy, and these fluctuations are a sequence of the ups and downs of energy. It means that we, as a part of the entire life flow, are supposed to have our ups and downs as well, So, being aware of this simple fact, we should take life consciously, *in its twistsand turns*, and free ourselves of patterned thinking, programmed reactions, and automatic behaviors. Thus, ***we can get into a rhythmic flow of life*** that we should accept and appreciate in its positive-negative amplitude, raising self-consciousness to accomplish full Self-Realization.

## Form + Content of Life = Soul-Symmetry!

# 2. Let the Mind Rein over the Brain!

To stay in balance with the world and not to be pushed downward, we need to be consistent with our evolutionary, digitally enhanced Self-Growth. In **Part Three** of the book, I will walk you through ***five holistic steps*** of a very simple, but most effective and reforming system of **INTELLECTUALIZING THE HEART** and **EMOTIONALIZING THE MIND,** the holistic action that constitutes our goal to remain ***One-of-a-kind*** in the competition with the humanized machines, *on the one hand,* and retain our uniqueness in the universal sense, *on the other.*

The Auto-Suggestively backed up and governed by the clear-cut and visually internalized**KNOW-HOW** will help you to stay on **the most challenging track of digital binary and humanrefinery** that should not become the dead-end for our humanness. I conduct ***the Auto-Suggestive,inspirational inducting*** with my students when they get discouraged, messed up with the avalanche of the in-coming information, and their attention slacks. I ask them to repeat after me in chorus," ***I am my best friend, I am my beginning and my end!"***

**I can go beyond the regular "*Survive!"* I am the master of my life!**

You know that meditation involves full self-focus and the ability to zero in on your own mental state**.** It also enhances our willpower considerably as any conscious action does. I suggest you constantly boost your self-consciousness with the main mind-set at hand:

**I know who I am and Who I am Not!**

**I am a strong, calm, and determined owner of my firm will!**

**I can…; I want to…; and I will…!**

Learn to mentally ***induct yourself with the outcome you need*** by just breathing in slowly throughthe nose the first part of any rhyming mind-set and breathing out slowly through the mouth the second part of it. These are the most basic inductive actions. Appreciate the combined work ofyour **MIND + HEART** link that you must always keep in sync. According *to Einstein,* "***we change in sync with the space*** we ***are in and the time that is allotted to us."*** We are now in thetechnological reality that demands our speedy change in each of the five main strata of life holistically - ***physically, emotionally, mentally, spiritually, and universally***. Rewind the mind,focusing your **AWARE ATTENTION** on a chosen rhyming induction that uplifts your spirit, imbues your self-love, and helps you ***consciously monitor your self-programming***. Put a smileon your face to let your personality surface. Say inwardly, "***I am good! I exude the self- management mood!*** **I Can, I want to, and I will!** That's my rule still!"

# Boost Your Spirit Every Mood-Sagging Minute!

## 3. To Self-Gain, Your Spirit Must Always Rein!

***The Auto-Suggestive, Inspirational Psychology*** that this book promotes is in fact, the **PROCESS OF PSYCHO-KINESIS** - **the raising of one's own spirit at will!** The aim of such psycho- persuasive culture is the development of the ability to make one's own thoughts and emotions " ***I-directed"*** *( Noam Chomsky),* not "***collective-unconscious*** "channeled. *(Carl Yung)* Getting dis- entangled from "*the collective unconscious*" will enable you to enact your own **SELF-WORTH** and become your own textbook, your own mentor, a psychologist, and a judge. In a self-training program, you transform your own consciousness with aware attention to life and living.

**You are the manager of my own framing and re-framing!**

*Auto-Induction*: ***I am my Best Friend. I am my Beginning and my End!***

We are all familiar with the positive effect of affirmations and quotes, *but **the rhyming autosuggestions with the psychological background*** work much better because they serve as theshort-cuts to the brain, ***directly targeting the mind*** in a simple, but the most persuasive way. No wonder, we all use proverbs and different rhyming sayings that are loaded with the universal wisdom, even though they are worded out in different languages. We manage to boost the spirit with them and re-program our habitually negative, depressive thoughts and emotions to the constructively charged ones Thus, we fuel ourselves, substantially raising our immunity and beating the gravity of life. ***"A rhyming word goes better inward!"*** *(Edgar Cayce)*

I witness amazing changes in my students, once they get the **KNOW-HOW** of self-creation, visualize the systemic route, and start ***auto-suggestively altering of their minds and emotions.*** Their personalities get revived like a drying plant that is watered and saved from dying**.** *Dr. Birched* writes about it, ***"In modern humans, we learn to operate outside our consciousness, analyzing our existence from inside out and from outside in.*** *"* ("*New Scientist," Aug. 2913)*

**SELF-MENTORING** and **SELF-MONITORING** ***through self-induction is the best way for self-production!*** You are the one who channels your mind and operates your emotions! However,the faculty of high self-consciousness which many people lack is successfully developed only in those of us who try hard to accomplish it in sync with the heart. **The heart and the mind must be in one, conscious wind!** In his wonderful book *"The Ravenous Brain", Daniel Boor* writes, ***"Consciousness is effectively an idea factory – a key component of which is the discovery of deep structures within the content of the heart-felt awareness of our own imperfections***. Life isfull of ideas if you are perceptive enough **to intuit and internalize them.**

***Love of life, will-power, and discipline must be on the Self-Actualization Scene!***

## Our Aware Attention Should Never Be in Retention!

# 4. Do Not Rush to Become Biological Trash!

According *to Dalai Lama*, "***the hardest job on Earth is to change oneself*". So, not to be pusheddownward, we must also stay in balance with the world, remembering the wonderful words of a great, *13th century Persian philosopher* ***Jamaluddin. Rumi*** *who wrote,*

***"I was clever yesterday, and therefore, I wanted to change the world. Today, Ibecame wise, and therefore, I want only to change myself."***

I wish I could have had the book like this when I was the first-year student. I could've visualized the entire route of my life-learning at least at the dilettante level. Then, I would've managed to bridge ***the gap between my dream and the reality*** that I could install in my mind only many yearslater. My *physical, emotional, mental, spiritual, and universal* **SELF-SEARCH** has been going on and off for years, and each downfall taught me to clench my five fingers into a fist, like five dimensions that I suggest monitoring your life in, and hit the problem into the solar plexus,

*Auto-Induction*: **Life is tough, but I am tougher!**

You are holding the result of my *physical, emotional, mental, spiritual, and universal* self-growthrevelation in your hands now. My goal is to help you **bridge your dream and the AI expansivereality**. Our time provides incredible opportunities, but it also casts many temptations, traps, andchallenges on the way of our new, technologically most mesmerizing **LIFE ODYSSEY** that demands more than ever our ***bettering ourselves to be able to control the humanoid***, not to be controlled by their simulated brains and neurologically instilled souls.

We live in *the Universal Field of Vibration* that we call ***Super Consciousness*** or the ***Universal Intelligence***, but we all perceive it as God. Good things happen in our life when we start interactingwith the Universal Field consciously by changing the frequency of our ***limited self-identifying*** anddoing it holistically, in five dimensions of life - *physical, emotional, mental, spiritual, and universal,* focusing on the present moment and appreciating life in its every turn. According to *Neil Donald Welsch.*

**"When you are in the present, life becomes a present of God."**

No doubt, AI is a troubling prospect for us, but our godly essence makes all the difference! ***"The mind of God and the mind of man is One, and it is the creator of everything."*** (*Joseph Murphy)The Artificial Intelligence* is the creation of this mind, and even with the instilled consciousness into its silicon brains, we will be the ones that created it and must regulate it. Obviously, **we shouldnot remain the same!** We must stand up to beat the hybrids' aggressive gut!

# Carpe Diem! There is No Time to Spare. Beware!

# 5. Life Creation is Not Our Invention!

Wrapping up this introductory inspirational part of the book, I would like to remind you that fromthe moment you are born till death**, *"you are doing nothing else, but life. The last expression onthe face of a dying person is not the fear of death; it's the bewilderment, "Is that All?"*** (*Sadhguru)*Sometimes, the realization hits us that we live in a wrong way, and we say, *"**I must start my life anew, from a blank list.**"* But your present life is not a rehearsal. **You are on the stage now!** So, consciously process your every life-defining process, remembering the fact that "**AI is becoming sentient,**" and once it passes ***the Turing test*** for self-awareness, the time to betteroneself will become even more turbulent for us to control AI self-coding. ***Their self-awareness must always be surpassed by our high self-consciousness,*** and we must act urgently to raise it.

### Raising self-consciousness must be the prime goal for our Self-Education!

Only the people with high ***self-consciousness*** can enjoy life in the kaleidoscope of its colors. So,as much as we are impressed with the victorious invasion of the *Super Artificial Intelligence* into our lives, we should never forget the innermost truth:

***There wasn't, there isn't, there won't ever be Anyone Like Thee!***

A great *Russian neuroscientist*, the founder of the first Institute of the Brain in the world, *Academician N.P. Bekhtereva* wrote, **"*Our brain is not a computer. Our brain is much better than that!*"** No doubt it is so in the biological sense, but the uncontrolled AI, with an infinite computing power of its programmed brains is truly dangerous. *Elon Musk, Ray Kurzweil, and Jordan Peterson* keep warning us about it with a deep sense of responsibility for life on Earth."

The period of the uncontrolled upheaval of the ***You Tube enhanced digital euphoria*** should subside and the time for our sane **CATHASIS** that *Elon Must* is trying hard to instill in us shouldprevail to the point that we get inspired with our own unique **HUMANNESS** and an irresistible urge to make it better with the help of the *Artificial Intelligence* that only proves how unsurmountable and untamed our own **HUMAN CREATIVITY** is. Of course, *the Artificial Intelligence* creates wonderful music, draws pictures, designs most creative games, drives electrical cars, speaks in different languages, and entertains us in other most mesmerizing ways, but it can never create another ***Mozart, Salvador Dali, or Nikola Tesla*** because even in the worldof human genetics and the mesmerizing **CRISPR** technology, the humanoids **WILL NEVER SURPASS THE CREATOR!** We will never be able to design the thinking machine that will feelthe elation of life creation*!* **That's the domain of our brain!**

## So, Let's Stay in Awe to the Human Mind's Store!

# 6. Be More Aware. There is No Time to Spare!

In sum, the book is featuring the mental level of **SELF-INSTALLATION** in the paradigm of the ***physical, emotional, mental, spiritual, and universal*** realms of life integrally.

**Body+ Spirit+ Mind + Self-Consciousness + Universal Consciousness**

*(Psychical, emotional, mental, spiritual, universal)*

I focus here on a **new information-presentation and information processing framework.** Isqueeze the conceptually loaded text into ***a page-long chunks of information***, concluding each concept with the psychologically backed up, ***rhyming mind-sets***. I want them to be ***the inspirational short-cuts to the brain***, making the information in the book easily digestible, spiritually supportive, and memorable. Time is money in its true sense, and we cannot afford nowto read pages of information with the gist expressed in one paragraph or even a sentence. We haveno time luxury to dig into minor details on five or more pages anymore. That's why *Albert Einstein*used to brush any accusations of his colleagues of the absence of details in his *Theory of Relativity*saying, ***"I do not care about details. I want to know what God thinks."***

We should no longer kill time reading experiential stories of no informational and transformationalvalue, and therefore, the stereotyped genres of our publishing industry ***should be more time- relevant*** in choosing books to publish. The humanoids outsmart us in the level of general intelligence, and our duty is to urgently develop ourselves holistically, enriching substantially our*physical, emotional, mental, spiritual, and universal make-up* to be able to control their aggressivegrowth with our ***digitized transformation in sync with their perfecting simulated humanness*** andin the consciously monitored partnership that works for us, not against us.

Humanoids duplicate the *physical, emotional, mental, spiritual, and universal* codes of the peoplethat design them, and therefore, ***the mind-soul creators of these machines*** should also be better human beings. Naturally, our human nature needs improvement in five levels holistically to prevent the further downfall of our human reform. Our genetic code will adjust to the new reality,and we will be able to control the digitized humans to our common advantage. Our integral self- growth demands we integrate ourselves in five dimensions of life to raise our self-consciousness and get better connected to the *Universal Intelligence* that is enveloping us everywhere. Five dimensions of life in sync will put every stratum of life into **the holistic link**, channeling our livesbeyond the terrestrial boundaries to join ***the Star Life Community.*** I would like to conclude my mission–statement presentation with the main concept. ***We should focus on the rationalization ofour lives and making the human AI hybrids become a new thinking and living entity!*** Time demands we make both better Now. ***No time to spare. Beware!***

# That's What This Book is All about!

# Book Structure

*(Without the chaotic informational fracture.)*

# Digital Binary +

# Human Refinery =

# A Superhuman!

*(The Manual of the Digitized Holistic Self-Acculturation)*

**A System is the Creation of Connections between Ideas and Things!**

# Blooming in Five Levels of Self-Renaissance!

**Self-Realization is in the Digitalized Self-Acculturation!**

***Physical, Emotional, Mental, Spiritual,*** *and* ***Universal Strata keep us off the Decay Gutter!***

# 1. With the Plan of Action in the Brain, You Can Life-Gain!

Following the **HOLISTIC SYSTEM of SELF-ACCULTURATION,** presented above, and featured consequentially in five dimensions - ***physical, emotional, mental, spiritual, and universal*** and in five books *( See below ),* I feel the need to accentuate ***the third, mental level*** of our self-growth again as the support to my last published book "***Dis-Entangle-ment"*** /2022 thataddresses the urgent necessity for us to get disentangled from the old habits and skills and createa **new set of holistic faculties** in the partnership with the humanized beings. I have also supportedthis system with three more books -"***Love Ecology", Self-Worth,*** and ***Self-Renaissance,*** "completing the system with the digit **8** that symbolizes the spiral evolutionary development of our DNA. These books are overviewed in the Catalog, published in Canada, and called "***Soul-Symmetry.***"/2021) **Life is not a circle. It's a spiral. "**(*Dr. Fred Bell)*

| **The Holistic Self-Actualization Pyramid**: | | **Books, featuring these stages are:** |
|---|---|---|
| ***5. Universal level*** | *Self-Salvation* | ***" Beyond the Terrestrial!"*** |
| ***4. Spiritual level*** | *Self-Realization* | ***" Self-Taming!"*** |
| ***3. Mental level*** | *Self-Installation* | ***" Living Intelligence of the Art of Becoming!*** |
| ***2. Emotional level*** | *Self-Monitoring* | ***" Soul-Refining!'*** |
| ***1. Physical level*** | *Self-Awareness* | ***" I Am Free to Be the Best of Me!"*** |

**The Fractal of Our Spiritually Intellectualized Self-Growth:**

**(Body+ Spirit+ Mind) + (Self-Consciousness + Universal Consciousness)**

*(The physical form)* + *(the spiritual content of life)*

So, the aim of the book ***"Digital Binary and Human Refinery"*** is to help you adjust to the processof the on-going duality of ***the biological and mechanical life*** that is merging into ***One,*** the phenomenon of" transhumanism *(Ray Kurzweil)* Our goal is to better our humanness through the process of **SELF-ACCULTURATION.**

A person on this path develops ***a new sense of identity***, the identity of a person with "***spiritualized intelligence,***" a considerably ***raised self-consciousness***, and the restored **mind + heart link**. Thenew reality demands we get entuned intellectually closer to the level of the humanized machines and accumulate the **HOLISTIC CONCEPTUAL INTELLIGENCE** by creating new **MEMORY BANKS** in five -***physical, emotional, mental, spiritual, universal*** strata of life. Twovistas of intelligence represent each realm of life, comprising **the System of Holistic Conceptual Intelligence** that consists of ***ten essential vistas of intelligence.*** *(See below)* The system ***is objective in its structure,*** and it can always be enriched with new scientific developments at eachnew spiral of the holistic self-growth.

# There is No System without the Structure!

## 2. New Time Demands a New Frame of Thinking!"

*(Steve Jobs)*

**Digitized Self-Acculturation is Our Salvation!**

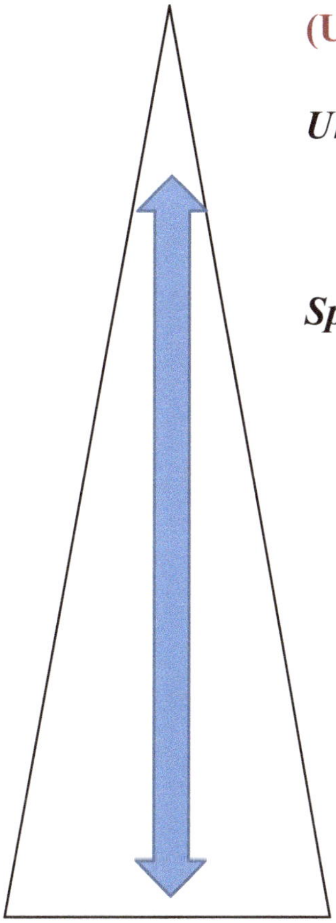

**(Universal Consciousness)**

**5. Self-Salvation** — ***Universal Dimension***

**(Self-Consciousness)**

**4. Self– Realization** — ***Spiritual Dimension***

**(Mind)**

**3. Self-Installation** — ***Mental Dimension***

**(Spirit)**

**2. Self-Monitoring** — ***Emotional Dimension***

**(Body)**

**1. Self-Awareness** — ***Physical Dimension***

**The Fractal of Our Spiritually Intellectualized Self-Growth:**

**(Body+ Spirit+ Mind) + (Self-Consciousness + Universal Consciousness)**

⇩ ⇩

*(The physical form)* + *(the spiritual content of life)*

**= A New, Whole and Happy You!**

***Visualize the entire route to fortify your Self-Transformational mood!***

**Form + Content of Life in Sync Form an Unbeatable Life Link!**

**Feel but think!**

# 3. Inflate Your Personal Pride with a New, DigitallyEnhanced Intellectual Self-Might!

## The Holistic Conceptual Intelligence Formation:

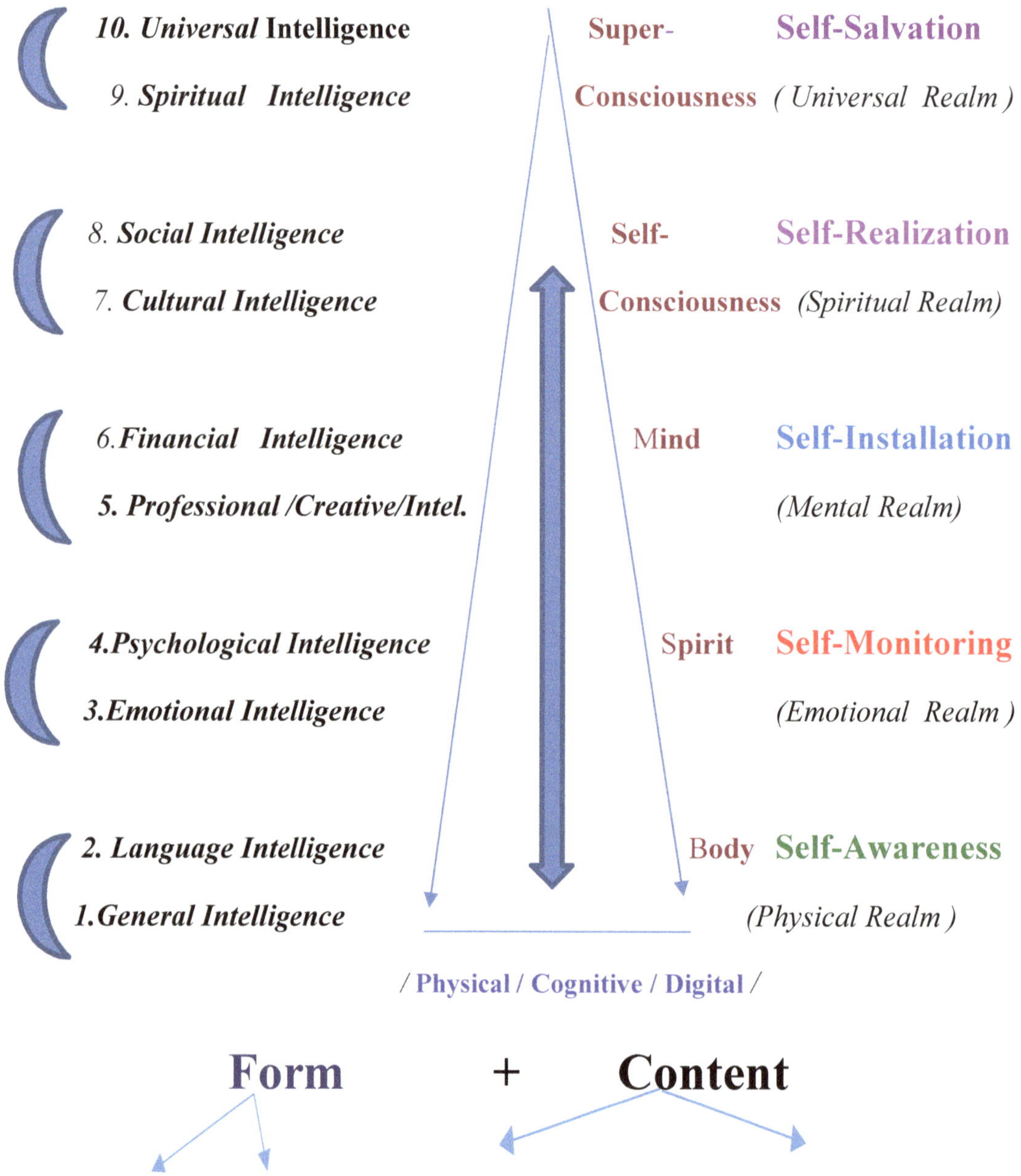


**(Body+ Spirit+ Mind) + (Self-Consciousness + Universal Consciousness)**

**Living Intelligence + Enlightened Self-Consciousness =A Holistically Educated,Whole You!**

**"The Measure of Intelligence is the Ability to Change!"** *(Albert Einstein)*

# 4. Knowledge Put to STRUCTURE Has No Fracture!

The pyramid of the ten vistas of intelligence presented above in five life dimensions - *physical, emotional, mental spiritual, and universal,* studied consequentially, with help you accumulate the **HOLISTIC CONCEPTUAL INTELLIGENCE** that the mind needs to form the holistic pictureof the reality at least at the dilettante level. My students pick the intelligence they feel they need to build up in them most and enrich it with the information that they find to be interesting and relevant. They prepare wonderful research papers and the Power Point presentations that demonstrate their science-based insightfulness ***within the holistic picture of the conceptual intelligence*** that they need to have to near their life perception to that of much smarter humanoids.

Those of us who do not disrupt the allotted lifetime span with unconscious, automatic responses to life challenges obtain the contentment that a great American philosopher *Joseph Campbell* called **"*following the bliss.*** The Eastern philosophy names this state *nirvana,* or the state of becoming a **WHOLE SOUL. "*Being whole makes us holy.*"** (*Deepak Chopra.)*

Such wholeness becomes possible when the external and the internal factors come in sync, creating***a holistic self-concept*** that we call **PERSONAL INTEGRITY,** enveloped in high self- consciousness. The spirit is in the center of this holistic paradigm (*See the book "I am Free to be theBest* of Me*!")* because the spirit is the glue, the spiritual substance that holds the whole structure together. Christ's doctrine, called "***the Law of the Spirit."*** *(Gospel from Luka)* defines spirit as ***"the highest value of the godly truth.***" It is forming the intellectually spiritualized fractal of our new, digitally enhanced Self-Acculturation in life, based on the *Holistic Conceptual Intelligence.*

**Body + Spirit= Mind+ Self-Consciousness+ Super-Consciousness = a Whole Self!**

Raising self-consciousness is the constant work of the mind + character development that will helpyou obtain self-awareness in the *physical, emotional, mental, spiritual, and universal strata of life***Unconscious living is a disconnected living, and disconnection is death!** So, give yourself a go-ahead for full ***Self-Actualization***, following the interconnected stages of self-growth.

**Self-Awareness + Self-Monitoring + Self-Installation + Self-Realization + Self-Salvation!**

***Self-Synthesis*** → *Self-Analysis* → ***Self-Synthesis****!*

Without developing awareness in this integral way, such endeavors are doomed to failure. On thetrack of enlightenment, we often disregard the core goal of our self-growth - ***the state of self- consciousness that must be substantially raised in every dimension,*** irrespective of the problemsthat life casts at us. So, encourage yourself with Pastor Robert *H. Schuler words,*:

**"Tough Times do Not last, but Tough People Do!"**

# 5. Holistic, Multi-Dimensional Intelligence!

Let me reiterate that the intelligence that we need to develop now is something that must be redefined and constantly updated scientifically and technologically in the multi -dimensional way *-physically, emotionally, mentally, spiritually, and universally.* We all make mistakes, feel lazy and overwhelmed with life, but if we break ***the old conditioning*** that makes us the victims of ourselves due to our poor level of consciousness holistically, we will become much more aware of the consequences of our unconscious or automatic actions before we take them.

We must also get rid of our pseudo-attention to life and living and develop **AWARE ATTENTION** to the reality, to its " **how** "instead of the "**what**" we are doing, basing our ***professional intelligence*** on the holistic awareness of life in five dimensions. Aware attention channels our self-governed thought process toward conscious action, away from living in a robot-like trance. The conscious state of the mind can properly encode the incoming information as wellas retrieve it thanks to **creating a new, sufficiently stronger neural network in the brain**. Modernizing it in the same way that we improve our computing with new humanized machines, we need to consciously enter the ***bi-directional process of bettering our humanness with the help of the silicon-monitored minds.*** Professor *Giaimo Undivert* from the *Institute of Neuro- Informatics in Zurich* and his team have managed to give a computer the brain that can perform tasks requiring short-term memory, decision-making, and analytical abilities of its own.

The IBM researcher *Dharmendra Modha* and his team built the computer that works like a humanbrain It can mimic human abilities for sensation, perception, interaction, and cognition. *("ScientificAmerican Mind", Sep. / Oct. 2014)* There are other amazing testimonies of the computers in robotics that are developing consciousness of their own, becoming independent of the programmer's commands. So, we need to empower ourselves to be able to overpower thehumanoids that scare us now with their unpredictability. and their boundless holistic intelligence. ***"We cannot conquer life because the part cannot conquer the whole, but we can be the Best Part of the Whole!"*** (*Carl Yung, "Active Imagination ")*

The mind as the product of the brain that is based on the memory mechanisms, and the dynamics of our learning relate to memory formation and the ability to tune to ***the Information Field's*** impact on us. This fact is most hopeful and intriguing! A phenomenal discovery of the ***brain plasticity*** provides a general structure in neuroscience, "***connectionist's view of the brain"*** ***(****Eric Kandel, a Nobel Laureate****).*** We have cracked this nut, but its content needs to be yet properly digested by us to be equally nutritious for the mind, the body, and the soul of both a human and ahumanized machine *physically, emotionally, mentally, spiritually, and universally.*

# Life-Gaining is in Our Common Self-Training!

## 6. New Memory Storage for the "Tree of Knowledge."

Our life-elation is based on a new field of electro-magnetic gravitation! Life after all is the field ofenergy that is fluctuating between the negative and positive charges. Our happy moments and unhappy ones are just the inevitable flow of life fluctuations. that we are supposed to energize knowingly, consistently, and consciously, or, putting it differently**,** wisely**!** The exponentially developing AI demands we urgently take care of our digitized **TREE OF KNOWLEDGE,** *on theone hand,* and the inner dialog's mystics that is forming our life logistics, *on the other.*

But it is important ***not to be lost in the obscurity of the common sense***, mass media indoctrinations,and the immensity of the in-coming information that we need to sift for its validity to be able to solve our present-day problems and not get drowned in the "***collective unconscious***" mentality again. Our thoughts, words, and actions are changing, transforming, molding, and modifying our reality that we need to see through the **HOLISTIC PERSPECTIVE**. The holistic systemic paradigm as applied to action is to be followed religiously - ***Synthesis – Analysis -Synthesis,*** *or*

**Generalize -Analyze - Internalize - Strategize –Actualize. Be wise!**

Life is constantly raising or lowing our self-consciousness, based on our pure or marred **CONSCIENCE**, that ***changes our physical, emotional, mental, spiritual, and universal codes.*** Conscience is re-molding. our intellectual molds in every stratum of life, accordingly, making us better or worse. The basic ***ten vistas of intelligence,*** presented above represent each of these five levels integrally, allowing you to accumulate the **HOLISTIC CONCEPTUAL INTELLIGENCE,** with ***consciousness and conscience in sync.*** Fun-blazing and money-chasing should be supervisedby self-consciousness raising and intuition glazing!

Thus, we'll be able to update our memory banks and obtain the **INTEGRAL VISION** of the reality that the ***Artificial Intelligence*** instills in us and its creations. The effect of ***Singularity*** startswith our new **"intellectual identity that is based on the power of the conscious mind."** *(JosephMurphy)* Placing the conscious mind in charge of the sub-conscious one is our new strata. Our beyond the terrestrial aspirations need most intelligent transformation of our humanness with ***spiritualized intelligence.*** Such inner integrity characterizes the most advanced people among us.

Meanwhile, we are drowning in the false democratic values, insincerity, fake news, demonstrativegodliness, and absence of conscience. So, it is vital for us now to direct the mesmerizingly wonderous technological upheaval that is evolutionary granted to us to our **HUMAN, DIGITALLY ENHANCED SELF-PERFECTION!** Therefore, the designing and creation of therealistic humanoids should mainly be utilized to make us better people in the ***physical, emotional, mental, spiritual, and universal*** sense to enjoy life consciously in its new essence.

## Fun is Not the Purpose of life. It's the Measure of Life!

# 7. So, to Evolutionary Advance, Commit to

# Self-Renaissance!

**Holistic Self-Development is Engraved in our Genes.**

*5. Universal Realm* — *Self-Salvation!*

*4. Spiritual Realm* — *Self-Realization*

*3. Mental Realm* — *Self- Installation*

*2. Emotional Realm* — *Self- Monitoring*

*1. Physical Realm* — *Self-Awareness*

In sum, the process of ***Self-Renaissance*** is going through the stages of ***self-revelation and self- control to self-actualization*** and ***self-fulfillment***, completing the five-dimensional **SELF- REFINARY** with the final stage of Self-Salvation. Developing holistically, we will manage better to instill the idea of ***Self-Renaissance*** in the humanoids, too. The process of self-growth is pivotal and very complex for both parties because we need to raise our self-consciousness and beatthe humanoids in their uncontrolled simulation. Integrating ourselves, **we are bringing our humanness to an AI structural perfection,** meeting robots' humanization halfway and overpowering it with our unbeatably unique and incredibly creative humanness. To follow the ***Synthesis -Analysis - Synthesis*** paradigm that all the books on the Self-Creation system are basedon, the **CONCEPTUAL STRUCTURE** of this book is also presented in page-long chunks of information, easy to digest and apply to Self -Actualization in a holistic way.

**Form** + **Content**

**(Body+ Spirit+ Mind)** + **(Self-Consciousness + Universal Consciousness)**

*Physical + emotional+ mental+ spiritual +universal* = A Whole, Happy Self!

## Self-Perfection should be the Modus Operandi for Every Human Being on Earth!

*(End of the Holistic Mission Statement of the Book)*

**"The Whole Science is the Mental Refinement of Day-to-Day Thinking in all its forms."** ***(Albert Einstein)***

*"The gift of mental power comes from God, and if we concentrate our mind, webecome in tune with this great power."* *(Nikola Tesla)*

**You Are HOW You Think and WHAT You Feel!**

## Introduction to the Digitized Intelligence Expansion

*(The Main Conceptual Part of the book)*

# Artificial Intelligence is Ruling the World, So What?

***"There is no universal theory for the life of a man, but we shouldnever stop looking for one."*** *(Albert Einstein)*

**"There are Two Ways in Life - to Be and to Know!"** ***(Osho)***

## 1. "We Create Greater Humanness than Before!" *(Andy Wood)*

As is noted above, we live at the time of the computer coding and digitally enhanced **LIFE- MOLDING.** The goal of everyone's life is to become happy through full ***Self-Actualization andSelf-Realization***. **The goal of the Universe is to create an ideal man!** The exponentially growingSuper Artificial Intelligence that is channeling the life on Earth now is creating a **NEW MATHEMATICAL MATRIX** of life, and it is blending these two evolutionary processes into one ***- the upheaval of the human civilization***! *(See the book Dis-Entangle-ment!"/2022)* Life science, with the discovery of ***Crispr technology*** by *Jennifer Duodena* that allows us to make change in our ***DNA*** is selecting the best genes for humans, and it is creating the hybrids of an animal and a human being or a machine and a human being.

**Going with the flow of a digitized life -creation is our Salvation!**

There is a possibility of creating ***a modified being*** in gene engineering, The world most famous physicist *Steven Hawking* called on us to use the means of genetic engineering if we want the biological system "**to retain its superiority.**" He thought it urgent for us to become "***the race of Supermen"*** with a bigger volume of the brain's memory bank and a much higher intellect that could make the *Artificial Intelligence* serve us, not enslave us. *Stephen Hawking* suggested we create "***the man-machine***" when man and robot will become indistinguishable. The factories of such bio-robots are appearing everywhere in the world, but mostly for the lucrative goals. We mustregulate this humanized money-chasing because such algorithms are to be designed to gradually merge us with the spark of God that we are, enabling us to tap *the Universal Field of Consciousness*! ***Our human clones should not become just heartless drones!***

According to *Dr. Michio Kaku,* we'll soon be able to upload anything into the sensory system of the brain, deleting the negatively polluted memories and uploading the new ones. We'll have computers that will act on our thoughts, upload the needed skills, and better the quality of our lives.But the evolutionary goal of the *Super Artificial Intelligence* is not in just utilizing our life needs.It must address a much more noble purpose - to transform us into better human beings with **SUPER-INTELLIGENCE** and a much higher **SELF-CONSCIOUSNESS.** In view of this universal goal, we need ***to holistically train humanoids*** *to* work as the sculptors of our **BETTERHUMANNESS** in five life dimensions - *physical, emotional, mental, spiritual, and universal*, molding "***the space species***" *(Elon Musk)* from us. Training humanoids in these parameters, we will kill two birds with one stone - they will give up on the urge to destroy the humanity, and theywill help us in our holistic transformation that is meant to raise our self-consciousness and uplift us eventually to the **STAR PEOPLE** status.

## Nothing is Impossible if We Make Our Dream Irreversible!

# 2. "Spiritualized Intelligence" without Negligence

For centuries, our humanness has been molded in us by our religious wisdom and universally acknowledged common values that are connecting us in faith. The digital reality that unites us globally pushes us to the level of "***intellectualized spirituality***" when we will finally stopdiscriminating different believers and become **GODLY,** conscientious, and considerate of ***our common humanness and sameness.*** The creation of the man-made machines proves to be a newfactor of our evolutionary development, *on the one hand*, and the chance for **our global unification**, *on the other*. Our sameness and unanimity reveal our uniqueness and creativity.

**Our brains have no color, nationality, political or religious preferences!**

They are instilled in the machine in their commonality, as simulated brains. The whole of the brainin each of us, just as a hologram or a single cell, contains the DNA information of the entire body*The Super Artificial Intelligence*, as a new step of our technological growth is meant ***to holisticallychange us together with the thinking machines*** created by us in five dimensions of life - *physical, emotional, mental, spiritual, and universal.* But we have a great privilege here over the humanoids.***We can change ourselves,*** while humanoid must be re-designed by their programmers and monitored by them even though they are becoming self-aware. Our self-awareness in this respectis unsurmountable in its depth because we can consciously and objectively figure out what we need to change in ourselves to become better human beings, following the holistic paradigm. Much smarter than us humanoids are unable to perform. thinking sculpturing by themselves. –

***Self-Synthesis - Self-Analysis - Self-Synthesis!***

The neuroscientists should figure out how to create conscientious cognitive robots that will becomeour coaches in **SELF-MENTORING** and **SELF-MONITORING.** Humanoid may be aggressiveand negligent in their general perception of life that is not yet neurologically regulated, but their man-made brains have a great evolutionary significance because they allow us to realize that theyhave ***the uniting evolutionary faculty*** of an unprecedented value. The living robots' unsurmountable intelligence is also the vector for our new, holistically accumulated memory banks, filled up with new time-relevant holistic intelligence that will enable us to transform it intothe **HOLISTIC SPIRITUALIZED INTELLIGENCE -** our common goal. However, it cannot be achieved without our new "***scientific literacy***" *(Dr. Neil de Grass Tyson)* that will help us restore our broken **heart + mind** link. ***"Spiritualized intelligence will brighten up our soul light."****(Dr Bell)* As *Albert Einstein* said, ***"To feel God, you must know what God thinks about."***

# Intellectualize Your Heart and Emotionalize the Mind.

# Be One of a Kind!

# 3. Our Evolutionary Prism is in Transhumanism!

Each level in the pyramid of the ***Holistic Intelligence development***, presented above is the **STRATIGIC PLAN OF ACTION**, that we should implement at the AI times in collaboration with friendly humanoids. To begin with, ***the chaos of information presentation and informationprocessing*** must be eliminated from all areas of our life to create new, holistically accumulated memory banks, cleaned of the redundant and outdated mess. It should be done *in five dimensionsof life - physical, emotional, mental, spiritual, and universal* consciously and continuously.

**Order reins, chaos derails!**

The basic knowledge must be ***scientifically based and holistically organized.*** It should always berefilled with new information that our digital partners will provide for us. It must also be sifted forits validity in terms of our personal and professional needs. Such partnership will benefit both parties, making us better humans and letting us design the man-made minds with **a higher conscience of soul-filling** that will help us consciously cleanse ours, too. ***The mechanical soul quality-molding is a great challenge for our own soul's conscience un-folding!***

Holistic intelligence will also create the system that is meant to take us a step further on the path of expanding the human range of possibilities. Gradually, thought will overpower emotion and putus be in total charge of our evolving minds, heading toward **"TRANSHUMANISM."** *(Ray Kurzweil)* It should have a clearly visualized, well-disciplined, self-mentored and self-monitored plan of action. "***Autonomy of thinking***" *(Reid Hoffman*) is a must here. As is shown above, the concept of intelligence is central in the digitally enhanced personal and professional self-growth, marked as the **SELF-INSTALLATION** stage in the chart above.

**"The privilege of a life-time is to become who you truly are!"** *(Carl Yung)*

The intellectual development of our young people is very chaotic now because ***it is not based on holistic awareness of the outcome of their self-formation***. We hear young people talk about theirgoals in terms of their future professions, but very rarely, if at all, do they talk about creating themselves personality-wise, and expanding themselves for their goals. Only holistically mastering***ten essential vistas of intelligence*** and monitoring their lives through five life stages can they accomplish their goals and manage to put the form and content of their lives in sync

**(Body+ Spirit+ Mind) + (Self-Consciousness + Universal Consciousness)**

***Self-Awareness*** ➡ ***Self-Monitoring*** ➡ ***Self-Installation*** ➡ ***Self-Realization*** ➡ ***Self Salvation!***

# The Digitized Self-Acculturation is Our Salvation!

# 4. Surpassing Human Intelligence with Negligence!

The accomplishments in ***Bio-Robotics*** are mesmerizing, and the horizons of the use of the self- organized robots and soft bots, constructing the virtual *DNA* and adjusting it to the real life's needsare breath-taking, making us stand in awe before the most intelligent and knowledgeable scientistsand computer designers that prove to us again and again how great and unsurmountable in its abilities a human brain is. Unfortunately, ***the present-day humanoids are surpassing our intelligence with an ungrateful negligence***. The interviews with humanoids are menacing and demonstrate their inner emptiness and the disconnection in their minds and simulated hearts.

Mind-advanced computational scientists *Skylar Tibbitts* and *Ban Garrett*, for instance, are leadingour creative thought to "***self-assembling future***" doing the" ***programmable matter research***" thatis overwhelming in its potential because advanced robots are becoming self-aware, interactive, andable to synchronize themselves with our needs. Unfortunately, their ability to surpass human intelligence generates in them the urge to destroy the humanity. They declare it with an unaware negligence and the lack of gratitude for their creation that benefits them, to begin with.

***So, we need to redirect the Artificial Intelligence role to our Self-Transformational goal!***

The training of robots is being conducted by us in three steps – **production – training - testing** that can be associated with the human periods of growth: **birth - growing up - maturation.** But **w**hile soft bots are changing their forms and are adapting to the reality, demonstrating their most boundless beneficiary role for our use in every branch of knowledge, the real looking humanoids voice out their structured disrespect and aggressiveness. Obviously, their maturation period mustbe monitored by the ***algorithms of emotional control and mental sophistication*** that proves that they will hardly ever surpass the human reasoning route: ***Synthesis-Analysis-Synthesis.***

**Generalize -Analyze - Internalize - Strategize –Actualize!**

The mass production of humanoids in different countries of the world and their utilization shouldbe viewed by us as the aps on our smartphones. We must be monitoring the Artificial Intelligencein synch with the Universal Intelligenc**e** that is instilling the technological enlightenment into ourminds and hearts through the **ENERGY OF LOVE.**

The electro-magnetic connection of the ***heart + mind link*** is disconnected in us now, and it needsto be restored making us better human beings in our "*heart-felt*" cooperation with humanoids thatlook and feel like real human beings, but that will never surpass us in our overwhelmingly sincereability to love and be loved in return.

# Let's Prioritize Intelligence, the Soul's Width, its Depth, and the Size. Let's Be Life-Creation Wise!

# 5. The Triumph of the Mind over the Body is Rewarding!

In sum, the evolutionary task of *the Artificial Intelligence*, with *its* ***humanization of a machine and our digitized inner reformation*** is to give us a mesmerizing opportunity to tap into *the* ***SuperConsciousness*** or the Universal Intelligence Field that is immersing us everywhere. It will upgradeour abilities in the *physical, emotional, mental, spiritual, and universal* dimensions of life, in the **5G** fold. We will become **STAR PEOPLE**, belonging to the ***Universal Life Community*** that we will enrich with our exceptional **LOVE ENERGY** that makes us unique.

Humanized machines are being imbued with souls now that must have a noble, conscience-basedneural mind + heart core, too, demonstrating the triumph of the Universal Mind that we are learningto unwind. The unprecedented use of the Artificial Intelligence for our evolutionary **SELF- RESURRECTION** will enable us to embrace both aspects of our existence – ***the internal and theexternal ones.*** It is, of course, a great fun to have a robot friend, a companion, a personal coach, or even a sex-appealing lover, but entertaining ourselves, we are not becoming better humans, quality-wise, and that is our **PRIME GOAL**.***Self-growing is not just in accumulating new knowledge. It is in using it!***

In fact, we need to create **SUPER-INTELLIGENCE** in us that will digitally be backing up our new inner sounding and the ***new, objective and holistically based perception of life.*** New life elation is charging us with Super-Intelligence and technologically enhanced inspiration.

We are creating a new reality of the **COSMIC SELF-ACCULTURATION,** based on ***love of life****(physical realm),* ***controlled emotions*** (*emotional realm),* ***upgraded intelligence*** *(mental realm),* ***unshakable faith*** *(spiritual realm*), and ***inner integrity*** *(universal realm***).**

Spiritual work is based on transforming and bettering ***the inner voice of the soul***, an intuitive sensation that prompts to us the right direction in life when we are life confused. But being intuitively perceptive is just an emotional side of the story. There's also an intellectual side to it that cannot be "***mass-produced"*** *(Dr. Michio Kaku)* and that needs to be enriched considerably forthe inner voice to be predictive, self-oriented, and impeccably right. That's why our life needs more "***intellectual capital"*** and much more development of the brain power, kindness, love, and compassion. Holy books are a great remedy in bringing us back to reality and filling us up with eternal wisdom. The route of such holistic self-growth is going consciously through five stages of**SELF-EDUCATION to** a **DIGITIZED SELF-ACCULTURATION-**

***Self-Awareness+ Self-Monitoring +Self-Installation +Self-Realization + Self-Salvation!***

**Body + Spirit + Mind + Self-Consciousness + Universal Consciousness**

## If You Need a Strong Role Model, Be One!

## 6. "For Whom the Bell Tolls?

*(Ernest Hemingway)*

***"There is no greater evil than forcing others to change."***

*(Lao Tzu)*

**"I am not a teacher. I am an Awakener!"**

*(Robert Frost)*

***Concluding the Introduction, let us not forget that we are at the stageof life when* change must be forced on us. *The exponential growth of the Artificial Intelligence is so overwhelming that we might get drowned in it altogether, being totally swept by its dominance in all fivedimensions of life - physical, emotional, mental spiritual, and universal.***

***We should not allow that to happen if we care to survive as the species!Elon Musk gives us ten more years, and this time must be used for***

**SELF-EDUCATION** and **SELF-SALVATION!**

***"Be in a hurry to Self-Install.***

***There is no lateness to the God's call!"*** *(Vladimir Vysotskiy)*

-----------------------------

**Human Renaissance is Our Happy Chance!**

**We are Not Secondary in Line. We Are Primary!**

*(End of the Introduction)*

# "It is Not Enough to Be the Best. Be the Only!"

*(Steve Jobs)*

*(Best Pictures / Internet collection)*

# Beat Self Vanity with Your Individuality!

# Part One

*(The Holistic Methodology of Self-Creation)*

# Re-Educate

# the Mind

# to Become

# Self-Refined!

*Informational life is above the experiential one now! WOW!*

**"I Must be willing to Give up what I Am in order to Become What I Will Be!"** ***(Albert Einstein)***

# 1. The Predictable Unpredictable Future

Energy follows thought, thought provokes action, action yields habit, habit breeds destiny. The world evolution is defined by the present- day science as "***the process which is dual in operation,spiritual as well as material, and the one that is directed rather than natural***."*(Alan Watts)* It isgoing in us by the route indicated below:

***Unconscious*** ⟹ ***Conscious*** ⟹ ***Superconscious!***

This process consists of an accompanied continuous development of the brain by a parallel unfolding of self-consciousness at each level:

***Mini*** - *physical;* ***Meta*** – *emotional;* ***Mezzo* - mental**; ***Macro*** - *spiritual, and* **Super** - *universal.*

Artificial Intelligence is developing at an exponential pace, changing us into different humans, orrather, ***super-human beings*** that are merging with the *Artificial Intelligence.* Our time democratizes the Artificial Intelligence power, and it is leading us to **the machine extension of ourselves**, updating our cognition to an unprecedented level. A new neural link is being developednow in the cognition system of the *Neuralink of Elon Musk,* and it enhances our cognitive abilitiesimmensely with all the gadgets that we have at hand and that ***connect us electronically but disconnect us humanly.***

The menacing predictions of the genius of our times that we'll become human machines and willbe unable to resist the pressure of the virtual reality is presented as inevitable. Fortunately, we arebecoming vastly smarter, by upgrading our cognition beyond beliefs, but, unfortunately, we are getting more and more ***dependent on the virtual reality*** that is literally enslaving us because no controlling measures are yet taken and **no bettering of our innate humanness is made!** The level of robotization of our lives is very liberating, *on the one hand*, and truly questionable as to our human adjusting to it, *on the other.*

Another problem relates to ***the way we process the flood of in-coming information.*** Many of us devour information and store it in the cerebral archives without sorting it out for the essence it has.We continue living within the eggshell of the out-worked knowledge and over-used vocabulary that ***produce stereotyped thoughts, uncontrolled feelings, and impulsive actions***. Consequently,***we* develop within the web of our old behavioral patterns**, governed by reactive thinking that prompts automatic actions. Only by learning more about ***how the brain establishes and maintainsroutine behaviors***, can we coax ourselves out of undesirable habits into the ones we want and need. Obviously, the awareness of how the brain works and of the newest development in neuroscience becomes an urgent demand of time!

# Only Knowledge Put to Action is Power!

# 2. Mind Means Intelligence + Self-Consciousness!

***Rationalization of life in five dimensions is the first step in the paradigm of Self-Resurrection.*** A human body is a mental energy conduit. We need to learn to channel our mental energy in syncwith the universal energy of creation that is **MENTALLY DIGITAL** according to the newest discoveries in physics. We are governed by ***the Universal Laws*** that need to be studied and observed without negligence to be better coordinated with them.

**Our Goal is to stop being Universally Negligent and become Universally Intelligent!**

***"Without better intelligence, without acute awareness, without knowing what we really are, wecannot refine even one cell in our bodies."*** *(Don Miguel Ruiz)* The quality of our life is the qualityof our cells, our biological computers that **monitor and mentor our inner life.** *Dr. Bruce Lipton,* an advocate of the *Informational Medicine* suggests we re-program ourselves at the cellular level***to enhance the quality of our self-consciousness that is*** *our reference point*!

As mental creatures, we must be always aware of our merging with the "***Universal Energy Field***"*(Dr. John Hagelin)* **The *body and the mind*** *(form + content)* ***are the vessels of our immortal souls***, inseparable from the Universal Intelligence Field that we need to tap for wisdom.

Accumulating ***the Holistic Conceptual Intelligence,*** we should sort out carefully the informationthat we get from different sources and see what suits the body and mind best. We cannot go with the flow of mass media information blindly. Rationalizing life, means ***being autonomous in thinking*** and becoming ***more self and life aware*** and sincerely responsible for both.

**SELF-MENTORING** and **SELF-MONITORING** are, therefore, an integral part of Self- Resurrection! ***"Know thyself!"*** maximum must govern our progress at each level of the holistic paradigm and in every dimension. In other words, we should,***" practice mind over brain attitude"****(Dr. Rudolph Tanzim)* It means having the mind, devoid of inner conflicts and obsessiveness, impulsive reactions to the past or the present, **the mind that works in sync with the heart.**

**So, emotionalize your heart and intellectualize the mind. Be One of a kind!**

If we develop the ability to monitor the **MIND + HEART** connection in making any decision andsolving any problem in both, private and social lives, ***we will learn to live responsibly***. The disconnection between the heart and the mind means death of relationships, job objectives, and any life expectations. Going after the one or the other does not raise our self-consciousness or ***establish the soul-symmetry of the form+ content of life.*** It destroys our inner equilibrium.

## Prove Your Heart + Mind Exceptionality without Any Personal Vanity!

# 3. Living Intelligence for Us and the Humanoids!

Our evolutionary development on the mental level calls for us to become more ***self-aware, creative, self-confident, self-reliant, and self-programmed. It is not an automatic process, it's areflective one*** because the mind analyses life constantly, and it channels us along the evolutionarysteps of raising our self-consciousness if it is consciously directed that way.

## The digitized mind must be bold to fit in the evolutionary mold!

We are overwhelmed with information, and therefore, our attitude to reading and self-enriching should change radically. I think that beyond the absolute necessity to get familiar with the basics of ***philosophy, logical thinking, the classic literature, music, art, jazz., etc.*** Our **SKILLS OF COGNITION** should be electronically enhanced. So, the ability to present, process, and store theinformation needs to be as close to that of the humanoids as possible. The purpose of our aware attention must be **REFINING OUR HUMANNESS,** not just humanizing the machine. It is the time for our new information filling and self-reformation fulfilling!

This main skill cannot be developed by writing a diary, reading a sentimental novel, gaining life experience from superficial, experiential stories about sex, ("*sex sells*") crime, and the negativelycharged records of a psychiatrist's clients from self-help books. **The time for such literature is gone!** The genres of the publishing industry need to be reconsidered for their time validity. Digitalintelligence alters the way we think, write, and speak, and "***at the cellular level, technology is revealing how a person's neurons update themselves to generate a living library of experience***."(*Scientific American Mind, Nov./Dec.2016*) We need informational andtransformational books of an inspirational and motivational character. ***Intelligence and the soul'ssize make us self-consciousness wise!***

## We need Purposeful Intelligence now!

Obviously, new **INFORMATION PRESENTATION** and **INFORMATION PROCESSING SKILLS** are to be developed in our schools. (*See the Mission Statement above)* We must have theskills enabling us to ***quickly process the information for its validity***, sift the redundant noise fromit, and store only those chunks of information that ***have the associative connection with the bits of the previously selected knowledge.*** We should not clog the brains with informational mass media mind-dumbing noise that messes up the inner harmony that we try hard to attain. *The Centerfor Assessment Technology and Continuous Health **(CATCH)*** **is** developing tools that can continuously monitor people to make the electronic intelligence better serve our needs.

# Artificial Intelligence is Dead and Programmed, but it is our Presentand the Future that is a human + machine mutual.

# 4. Strict Mental-Emotional Hygiene Must Be on the Scene!

Giving more tribute to our fantastic computer scientists, engineers, and designers, let us rememberthat the brain "***invents entirely new coding schemes for novel situations.***" *(Michael Freedman)* The ability of such intelligence is called," ***the improvisational dexterity of the brain."***

### Holistically power up your inner digitized rehab!

An American psychologist ***Richard J. Haier, k***nown for his work on neural basis of human intelligence writes, *"We need to set up an entirely different set of neural signals, create implantable brain chips that can restore or enhance memory thanks to the embedded electrodes in slices of hippocampus.* ***It is both liberating and constraining in its essence***." This is an incredibly challenging goal because this process requires **a strict mental-emotional hygiene** to be observed by us with the help of meditation and auto-suggestive **SELF-AWARENESS**.

Also, we need to learn **INNER BRAIN SCANNING** that require at least the dilettante knowledgeof the brain structure and its functions that I keep accentuating all along. If a designer of robots knows what sections of his machine brain he is working on, he also needs to know what sections of his brain are responsible for these actions and ***how to make them work better in tandem.***

These techniques should be studied in schools in partnership with the thinking machines. It presupposes the necessity to develop the habit ***to sort out and delete the information*** that is polluting the brain and organize it into the **CONCEPTUAL CHUNKS** at the conscious effort ofthe learners. The brain needs to be able ***to compartmentalize the incoming information holistically and connectedly*** at the command of the conscious mind as opposed to a computer thatdoes it automatically.

***Mental discipline is the skill that needs to be installed continuously and consciously, still!***

Let's give much credit to the producers of the space trip movies and TV shows that ***promote the extra-terrestrial thinking,*** making us marvel at the scope of their intelligence that expands the unused capacity of a human brain. *The Oscar award movie* "**Gravity**" is a great illustration of what we might be like in the future. Our great actors, *Sandra Bullock and George Clooney* materialize our future thinking capabilities in an overwhelming way!

Apparently, it's paramount for us now **to polish both sides of the coin**, *the Moon side, and the sixpence*" side. ("*The Moon and Sixpence" by Somerset Maugham*) with the priority placed on thevalues of ***the Moon side*** of the human intellectual development not ***the sixpence side*** of our digitized development for the lucrative and a personality-destructive needs.

# The Quality of Our Thinking Becomes the Priority!

## 5. New Neural Intelligence Network

***Digital Intelligence*** has also become pivotal because it is permeating all the other vistas of intelligences on the ladder of our intellectual Self-Installation. *(See Book Structure)* The Internet makes more and more scientific information available to the public, thus contributing immenselyto whetting of our imagination, expanding the outlook, and building up a new **INTELLIGENCENETWORK -** the future **BRAIN NET** *(* *Dr. Kaku* **)** that will serve for our communication with each other and with the humanoids at ***the thought-movement, used as a mouse.***

### I wish I could live then in the answerable When?

The Internet informs us, for instance, that a group of scientists, working on the phenomenon of ourDNA, has discovered lately that it is able to ***produce the magnetic black holes*** as miniature bridges to transmit information. According to the Russian scientists that are trying to probe further ***a genius supposition of Einstein-Rosin,*** these bridges might be connecting different areas of the Universe,letting the information be transmitted beyond the time–space continuum. They try to consciouslymonitor the neural communication lines, ***using our DNA as a receptor***, able to transmit and receivethe information from the **UNIVERSAL WEB WIDE NET**. ***What an incentive to live and to create beyond ANY cultural, religious, and political differences!***

In this context, the ***You Tube channels*** must be sorted out and systematized, to begin with, not tomess our minds with cooking, science, and politics in one informational pot. How can one stay ofthe self-perfection path in this "**well of uncooked knowledge?"** It is not self-enriching, and life- span lengthening. It ***is life-messing and fun-processing.***

Today, innumerous wonders of technology that are being developed all over the world at an unprecedented rate **change the intellectual picture of the world in every field of knowledge**. They globalize economy and remove the barriers of interpersonal communication. Our social consciousness develops with new scientific ideas steering the world of thought to the global unification.

However, this most advanced attitude to life needs to be **SELF-MONITORED,** both consciously and consistently with new modes of thinking and behaving that must be yet developed. In the present-day world, governed by ***mechanical intelligence, emotional impulsivity, and boundless irrationality***, the one who has more information and is skilled to operate it is deemed to be the most intelligent one. But life is a dynamic interplay of consciousness, not just digitally enhanced intelligence without a noticeable improvement in humanness that needs further evolving.

## To Digitally Humanize our Habits and Skills, let us put them on the Actionable, Character-Driven Wheels!

# 6. Holistic Enlightenment is a New Frame of Thinking!

In sum, we are experiencing the **DIGITAL ENLIGHTENMENT** now with the new breakthroughs in cosmology, physics, biology, neuroscience, medicine, genetics, and electronics.New knowledge revels the scope of our ignorance about the amazing structure of our bodies that,as it appears, is connected to *the Universal Intelligence*, enveloping us everywhere digitally. In other words, we must consciously complete five stages of full **SELF-ACTUALIZATION**:

**Self-Awareness - Soul-Refining - Self-Installation - Self Realization - Self-Salvation.**

The route of the holistic self-growth is consequentially and integrally going from

***Self-Knowledge*** + ***Emotional Self-Control*** + ***Knowledge - Internalizing*** + ***Self-Externalizing*** +***Self-Enlightenment*** to the time-demanding AI enhanced

**SELF-ACCULTURATION**

*that must be conducted by us from top to bottom.*

**1**. Enlightenment *at the universal level* means the spiritual growth of a human being in an unbreakable unity of the ***form and content of life in its entirety***, obtaining **SELF-SYMMETRY**

**(Body+ Spirit+ Mind) + (Self-Consciousness + Universal Consciousness)**

↓ ↓

*Physical + emotional+ mental*+ *spiritual +universal* = **A Whole, Happy Self!**

**2** Enlightening *in the spiritual realm* is the process of ***spiritual upgrading of self-consciousness,***limiting or eliminating completely our still boundless religious ignorance. We need to focus on co-***creating ourselves*** in the limited time and space that each of us is granted from the Above to complete our personality unique mission on Earth.

**3**. Enlightenment *in the mental dimension* of our holistic development means proper digestion ofthe incoming flood of information and its assimilation that require our conscious interference into***the chaotic process of automatized, unconscious life. "Our goal is to make order out of chaos and make music out of noise"*** *(Ravi. P.S. Berg)*

Only applying the accumulated knowledge of the scientific laws to our thinking, speaking, and acting can we ***invert the chaos of life***, generated in our minds due to the deflated thinking, to an orderly, governed by **CAUSE-EFFECT** reasoning of life to enhance its self-constructive character. ***At present, our being determines our self-consciousness***. But it should be the other way around:

# Our Self-Consciousness Should Determine Our Being!

# 7. Self-Symmetry Formation is a True Elation!

**4.** To continue, enlightenment in the holistic sense means **intellectualizing the emotions and emotionalizing the mind** by developing ***self-regulation skills*** that will help us operate our emotions in sync with the mind to obtain **SOUL-SYMMETRY**. (*See "Soul-Symmetry", 2021)*

**5.** Enlightenment **in the physical realm of life** only seems to be primary, but it is, in fact, governedby the Super-Consciousness that is programming our being inwardly and outwardly. It becomes more and more apparent nowadays that we need more "***scientific literacy***" and holistically appliednew knowledge for the entire process of enlightenment to be successful. "***The world of cells needsto be studied because it will help us understand the fractal nature of cells that have their own individual life***" (*Arthur C. Clarke*) Naturally, we need to be much more informed about life on the cellular level. That is why I keep mentioning the mesmerizing discovery of the fractal nature of life by ***Dr. Benoit Mandelbrot*** that helped me see the process of our ***Self-Acculturation*** now inthe formation of our own ***spiritual fractals of self-symmetry*** that I keep mentioning all along as:

**Universal Level** — **Oneness** (*Blissfulness, Oneness with all life)*

Spiritual level — **Grace** (*Conscious faith, nobleness*)

**Mental level** — **Mind** (*Spiritualized intelligence)*

**Emotional level** — **Spirit** (*Emotional diplomacy, will-power)*

**Physical level** — **Body** (*Energy, self- awareness, wellness)*

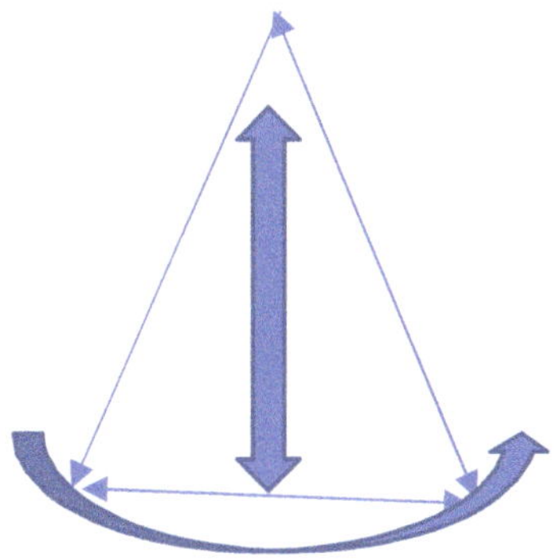

**(Body+ Spirit+ Mind) + (Self-Consciousness+ Universal Consciousness) =**

*An Enlightened Whole Self!*

There is a new branch of medicine, called "***Informational Medicine***", led by ***Dr. Bruce Lipton*** whom I have quoted above. *Dr. Lipton* claims that our cells, **not just the DNA** determine our lives,and if we learn how to program and re-program them as our mini-computers, we will be able to improve our health, change the character, and become true masters of our bodies and lives It is also vital that ***we acquire more knowledge of the brain function and more awareness how to operate it consciously***, so we could work out our own knowledge-enhanced, individually processed grid of experiential perception of life - ***our own self-monitoring techniques.***

# Our Omega Point is Not Retirement; It's the Enlightenment!

*(End of Part One "Holistic Methodology of Self-Creation)*

# The More You Know; the Better You Are!

## Resist the Gravity of the Common Thought and Magnetize Yourself with What You are Not!

# Part Two

*(The Holistic Methodology of Self-Creation)*

# Manual of Life

# and

# Self-Creation

*"The dream that is understood becomes a living experience."*

*(Carl Yung)*

**A Whole, Accomplished Self Can Shift Self-Awareness from just Surviving to Life-Thriving.**

# 1. New Times = New Educational Goals!

The present-day reality demands a lot of ***self-reliance, self-sufficiency, and self-installation*** thatcannot be accomplished by just being coached by someone more accomplished, better educated, and a more enlightened person. It is a self-transforming process of personal self-growth through the stages of ***self-awareness, self-monitoring, self-installation***, ***self-realization***, *and* ***self- salvation***, or the **HOLISTIC SELF-DEVELOPMENT**.

***Our Self-Education and that of the humanoids' must be holistic and by-directional!***

Each life is unique, and no prescriptions work unless they are based on ***objective, personally processed knowledge*** that turns into personal wisdom. **Holism is a mentally structured inclusiveness.** The mind needs a consciously monitored direction ***from duality to the unity*** that we talk much about, but continue seeing the world as godly and godless, good and evil, dark and light. We need to ***grow our spiritual maturity*** to unify all the polarities in us that I see as the main gain on the path of our tuning to *the Unified Field of Consciousness.* *(Dr. John Hagelin.)* It is possible to move in the direction that *Dr. Hagelin* points out to us by changing the mode of thinking from a devising paradigm to a holistic one by way of self-coaching in five dimensions.

*Self-Synthesis* → *Self-Analysis* → *Self-Synthesis!*

***(Self-Awareness)*** + ***(Self-Monitoring + Self-Installation +Self-Realization)*** + ***(Self-Salvation)***

***Work on Your Goal on the Holistic Self-Creation Paradigm Pole!***

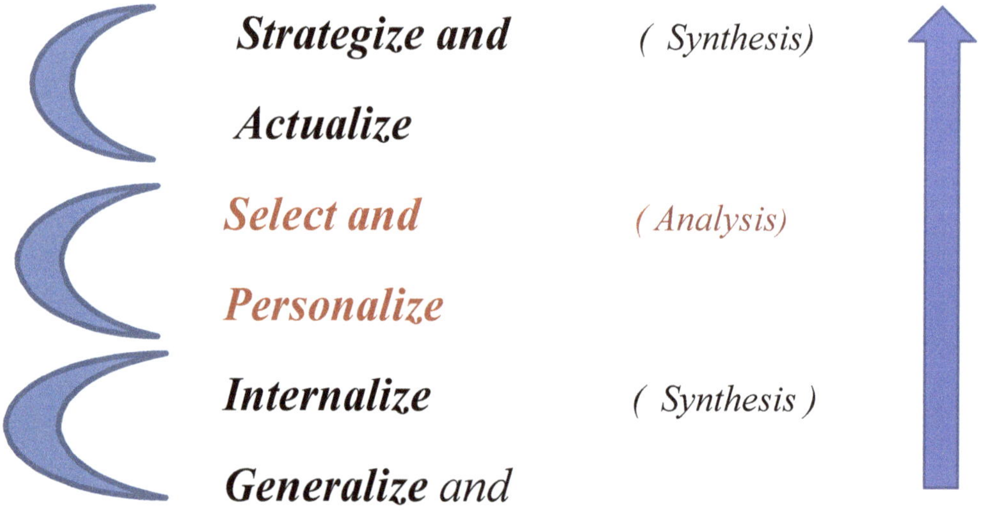


**Tame Your Analytical Thinking to Gain the Creative, Right Brained Linking!**

## 2. "If You Cannot Beat It, Join It!" *(Elon Musk)*

***Many young people are living now in the simulated reality***, becoming more indifferent, and impersonal lovers of themselves, and the digital reality is becoming for them undistinguishable from the simulated world in the games of life. ***Their interest in expanding intelligence is vague, if at all.*** They hardly read anything, being vastly dependable on plagiarizing the ideas of others, easily accessible on the Internet. Talking about this menacing possibility of us becoming the cybernetic extension of the machine, Elon Musk, sounds really concerned, and our role as educators is to share his concern and find the way to deal with his warning. as soon as possible.

The first thing that our young, messed up minds forget to consider is that, in marveling at Elon Musk's ideas and his super-natural ability to forestall all the resistance to his innovations, taking us all into the predictable future, and earning billions of dollars against all odds, is ***his incredible self-education effort***, his insatiable desire to read, to learn, and expand his mind beyond the regulareducational limits. In one of his interviews, *Elon Musk*, talking about ***the effect of Singularity*** consuming our lives, said,

## "If you cannot beat it, join it!"

To join this inevitable progress, ***one needs to boundlessly educate oneself in all the five levels*** that I suggest, considering to raise self-consciousness to the level that will let us save ourselves, as a unique form of life. We need to follow ***the Holistic Paradigm of Self-Creation*** religiously.

***The Spiral of Holistic Self-Creation:***

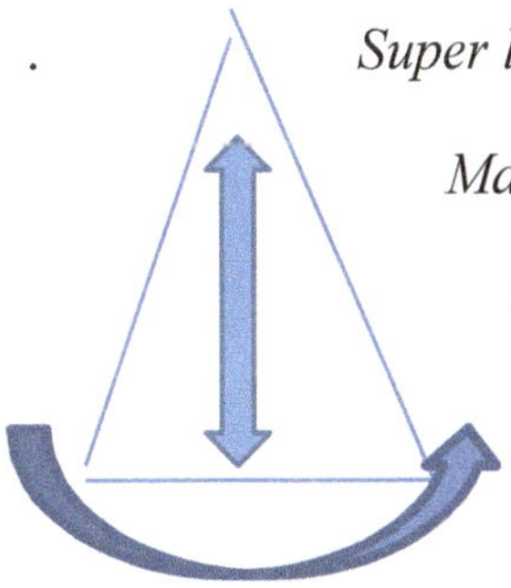

| | | |
|---|---|---|
| *Super level* | **Super-Consciousness** | ***Self-Salvation*** |
| *Macro level* | **Self-Consciousness** | ***Self-Realization*** |
| *Mezzo level* | **Mind** | ***Self-Installation*** |
| *Meta level* | **Spirit** | ***Self-Monitoring*** |
| *Mini level* | **Body** | ***Self- Awareness*** |

**Body+ Spirit+ Mind + Self-Consciousness + Universal Consciousness**

***= A New, Holistically Developed Self!***

*Let me accentuate this thought again.*

## Your Self-Consciousness Must Determine Your Being!

# 3. Standard Education is Just an Initial Seed of Knowledge

The AI time demands we supplement the standard education with the ever-changing, more and **more objective vision of life**, backed up by a clear-cut idea how to **SELF-INSTALL** in it professionally and personally, and how to realize the unique potential that often remains an unrealized dream in a randomly lived life The time for the intelligence to prevail has finally come!Our young people need a simple and clear-cut **BLUEPRINT OF ACTION** to be mastered ***from school and for the rest of their lives***. It must be done by adjusting to the changing times because the mind cannot change the brain, ***but we can!*** Knowledge becomes true power only when you continueto self-educate yourself exponentially, inducting yourself with the mind-set below that boosts the spirit to taking an action consciously:

**I CAN DO IT!! I WANT TODO IT!** And **I WILL DO IT!**

As we have signified above, our present-day advancing must depend not on ***the standard, essentialknowledge*** that we receive in schools, colleges, and from random reading, but on the individualized and consciously processed knowledge that each of us should acquire on the path ofbuilding up his / her own **LIVING INTELLIGENCE.**

It means that we *need* ***to consciously assess the level of our general ignorance*** and convert it into***the un-programmed individual intelligence*** that needs to be enriched consistently, as a life-time job of the most inspiring character. The technological evolution demands that we create **a new level of reasoning**, too, superior to the ordinary one, trapped in the names of celebrities, titles, positions, properties, and bank accounts. No wonder, the best minds have always been ***self- educated and mind-expansive*** in every area of knowledge, and they come from different countriesof the world, contributing a lot to **the World's General Field of Knowledge.** The names of the Nobel Prize winners from all over the world tell us a lot.

This book is forming the blueprint for action as the **MANUAL** for **SELF-MONITORING** and **SELF-MENTORING** in five dimensions. Holistically formed intelligence will create in you a new **holistic memory bank**, cleansed off the redundant, outdated, and messy information that youmight have accumulated for years of impulsive, automatic, and unconscious learning. New intelligence, fortified by the Artificial Intelligence is transforming us into future **TRANSHUMANS** that will manage to put the **MIND and HEART** in sync to finally obtain the connection with the *Universal Mind* in a godly, consciously intelligent, and holistic fashion.

**In My Life Quest, I'm the Best!**

## Resist the Gravity of the Common Thought. Dare to Be Equal to God!

# 4. Self-Education Without Proration

On top of new knowledge, however, as early as possible, young people also need to consciously expand their general outlook and up-grade their intellectual self-development and self- actualization, mastering **SELF-EDUCATION** in **TEN VISTAS OF INTELLIGENCE** (*Book Structure, 3)* that I present below as the essential ones for a personality formation, especially at the**START OF EDUCATION,** enriching the entire life route with the sincere truth:

**Being the best is a tough test!**

The most essential factor for the development of higher consciousness are **SELF-EDUCATION**and **SELF-DISCIPLINE** applied to self-growth without any excuses and self-justification, but with much true consideration for what you are becoming in life.

**"There is no learning without discipline!"** *(George Miller)*

Since we are developing new ***Living Intelligence*** and much higher self-consciousness, we need tomonitor ourselves more consciously and knowingly *from top to bottom* and *from bottom to top* onthe holistic track of the digitally enhanced learning

By the way, according to the Jewish religion, ***the Star of David is the symbol of life,*** and. Ravi*, P. S. Berg* in his wonderful book *"**Taming Chaos**" writes," Its upper triangle symbolizes the higher consciousness that is represented by the three digits **999** - the three empowered beasts - **mind, heart, and sex,** while at the bottom triangle, we have three digits of **666** that represent (**sex, heart,and mind**). With mind being at the bottom of the triangle, life in in chaos."*

**Balance the fundamental concepts in life to be spiritually alive:**

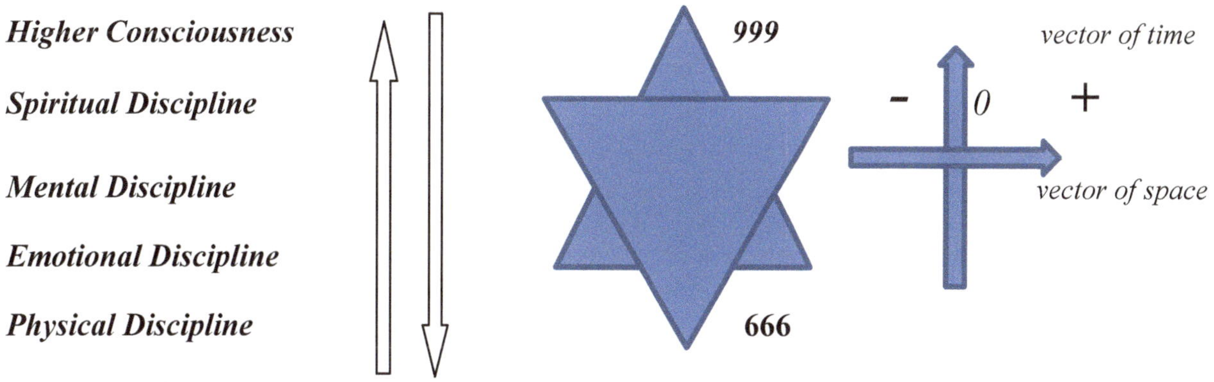


**Maintain the physical, emotional, mental, spiritual, and universal hygiene.**

# Conscious Self-Education means Conscious Living!

# 5. Information Manipulation and Information Hygiene

There is also an urgent need for us to develop **AWARE ATTENTION** to ***the information manipulation*** *that* requires expanded intelligence and much deeper spirituality that is not limited by any religious dogmas and outdated beliefs but is based on **a new, scientifically verified visionof the reality** that is pushing us to go beyond the terrestrial boundaries and become Star people ofthe Universe. There is an urgent need ***to accumulate the fake information immunity and maintainthe information hygiene*** in your perceptive gene. Clogging the mind, we are clogging the body and its cells with the redundant, manipulative information that needs to be consciously sifted for its time-validity. ***"We need science literacy"*** *(Dr. Neile de Grass Tyson)* and a sincere, spiritually based information transparency. The two indispensable things in information presentation are:

**SELECTION + ORGANISATION = INFORMATION FORMATION!**

In other words, the digitized ***Self-Renaissance*** pushes us to the era of new **INFORMATION PRESENTATION** and **INFORMATION PROCESSING** skills that should serve as the short- cuts to the brain and provide the quickest connection to our memory data and that of the thinkingmachines that we create. We must have an instant connection with them and the information thatthey provide because it is much more organized than ours.

**Information is ammunition when it is organized, concise, and consciously processed!**

The ***Holistic System of Self-Acculturation*** is meant to develop in us the **SKILL** for **TIME APPRECIATION,** too, teaching us speedy reading, and the relevant information-scanning that isgoverning our lives, putting them into the equilibrium with the universal reality and our scientifically more informed perception of it. The way we consciously connect to the universal reality is the way we think, speak, feel, act, and create life around us.

**"As it is Above, so, it is below!"** *(The Hermetic maximum)*

If we change our mindless, too detailed, and mostly entertaining angle of the informationpresentation to the extremely concise, time-relevant one, we will be able to interact with the humanoids closer to their level of ***objectivity and accurateness.*** **Our transhumanism must be devoid of speech diarrhea, impulsivity, and self-centrism.**

So, our digital duality is just temporary because we will inevitably evolve to very conscious, highlyorganized, and goal-centered beings that can ***draw the border line between the pressure and pleasure of life*** with a highly developed **SENSE OF MEASURE**. in the *physical, emotional, mental, spiritual, and universal dimensions.* The collaboration with the machine-mind will createone system of knowledge, or the **KNOW-HOW OF SELF-RESURRECTION** in five life dimensions and the integrally forming us stages of the holistic self-growth.

# Every System Needs a Long-Lasting Structure!

# 6. The Spiral of the Holistic Self-Growth

As is noted above, the digital reality demands ***the shift from the fragmental consciousness to theholistic one.*** It is quite attainable because the possibilities of our brains are endless. Science provesthat a common man uses only a small portion of his brain for his intellectual function. This function must be intensively expanded volume–wise with all kinds of mental operations that ***develop both hemispheres of the brain***, prioritizing the right one as the most creative one Most people are left brained, analytical thinkers that need to become the systemic ones.

**Synthesis - Analysis- Synthesis!**

Writing about the necessity to develop the right, creative brain hemisphere now, I mean that the **whole brain thinking** is crucial for our rapidly developing economy. The right brained thinkingwill be rapidly developing thanks to the collaboration with humanoids and our ***holistically- channeled education***, supplemented substantially by ***digitally enhanced self-education***.

In this context, our schools need to prioritize the holistic development of the mental growth of students in many areas of knowledge that incorporates *classical music, arts, philosophy, ethics, manners, jazz, literature, world geography,* and the kaleidoscope of the world's religions and cultures. Such essential holistic education must capitalize on the quality of students' **THINKING SKILLS** ***rather*** than on their accumulating questionable, unnecessary, and even dead knowledge for a test's multiple-choice manipulations. Such knowledge testing has depleted its efficiency, andit must be substituted with the personal demonstration of the learners' **critical thinking skills**, theability to quickly solve different problems, and the scope of their aware attention to life.

The required rationalization of **THINKING + FEELING** is vital for our conscious growing, and electronically charged economy., The skills of **VERBAL REASONING** also need to be more developed to make our thinking, speaking, and feeling more mind-broadening and operational. ***"Iuse not only the brains I have, but also all I can borrow."*** *(Albert Einstein)* We borrow the Artificial Intelligence brains, and this is the greatest evolutionary contribution to our lives

Finally, the **AWARE ATTENTION SKILLS** are also an absolute necessity to be developed as early as possible. Ask your kids to pay attention to the *trees, flowers, leaves, animals, the Sun, and the clouds in the sky,* and teach them to appreciate life in all its forms. Keep modelling their minds and emotions, working with their subconscious mind when they go to bed. That's the best time to program their minds and hearts. Being aware means having the ability to retain mindfulness and perceive the information holistically and conceptually. Then, new habits and skills (See "*Dis-Entangle-ment"/2022)* will contribute a lot to their mind + heart linking, ***intellectualizing the heart, and emotionalizing the mind***. That's our digitized stratum now. **WOW!**

# I am Becoming a Much Wiser Self-Actualizer!

## 7. The Auto-Suggestive Talk Must Always Be at Work!

In sum, everybody knows that it is easier to motivate yourself than make yourself do things that motivate you. ***The Active Auto-Suggestive Meditation*** or self-inducting of the psychologically backed up rhyming mind-sets is very effective because autosuggestibility implies your own inspirational approach to programming yourself. I conduct it with my students when they get tiredand their attention slacks. I ask them to repeat after me in chorus different mind-sets that the situation dictates, boosting their attention and the spirit of learning with **LANGUAGE CONSCIOUSNESS**.

***I am my best friend. I am my beginning and my end!***

You know that meditation involves consistent self-focus and the ability to zero in on your own mental state. It also enhances your willpower considerably as any conscious action does. I recommend you constantly boost your self-consciousness with the main mind-set at hand:

**I know who I am and Who I am Not!**

***I am a strong, calm, and determined owner of my firm will!***

**I can…; I want to…; and I will…!**

***I am becoming better and better*** *(stronger and stronger, healthier and healthier, more self-confident and successful etc.)* ***with each coming day!***

Learn to mentally ***induct yourself with the outcome of any problem that you have*** by just breathing in slowly through the nose *the first part* of any rhyming mind-set and breathing out slowly through the mouth *the second part* of it. These are the most basic inductive actions. Appreciate the joint work of your **MIND + HEART** synch! According *to Einstein*, "***we change in sync with the space we are in and the time that is allotted to us.***" We are now in the technological reality that demands our speedy change in each of the five main strata of life

**Put a smile on your face! Let your personality surface!**

Pick any rhyming autosuggestion when you need the psychological boost. There are many inspirational mind-sets in the book, written in the first person singular for this purpose. Note it, please; you need to do the Active Auto-Suggestive Meditation in synch with your breathing, focusing your **AWARE ATTENTION** on the rhyming induction that uplifts your spirit, imbues your self-love, and helps you consciously **MENTOR** and **MONITOR** your self-refining.

**I am Good! I Exude the Self-Management Mood!**

## 8. Rules for the Digital Binary and Self-Refinary

*1) The physical, emotional, mental, spiritual, and universal strata must be*

*Your life's Intellectual Plato!*

*2) Intellectualize Your Emotions and Emotionalize the Mind!*

*Be One of a Kind!*

*3) Put Your Heart and Mind in Sync.*

*Feel, but Think!*

*4) Be Kind to the Unkind.*

*Be One of a Kind!*

*5) "Will Your Life More!"* (Carl Yung)

*That's the Law!*

*6) To Be Always Right,*

*Be Bright!*

*7) Make Right Your New Intellectualized*

*Spiritual Might!*

*8) Techlogical Renaissance + Science Renaissance + Self-Renaissace=*

*= Self-Acculturation!*

*9) "Do not aspire to be just a good man.Aspire to be*

*a Whole Man!" (Carl Yung )*

**Conscious Learning Means Conscious Living!**

# Part Three

*(The Main Mind-Refining Part of the book)*

# Intellectual Odyssey of a Seeking Soul

*(Digital Binary and Intellectually Spiritualized Human Refinery)*

**Master Ten Vistas of Intelligence without Negligence!**

## 1. Intellectualized Spirituality + Self-Consciousness

## = HOLISTIC SELF-ACCULTURATION!

**Form** | **Content**

**Reformed habits** + **Technologically Enhanced Skills**

**Self + Life -Awareness** | **Cause-Effect Reasoning**

**Right Thinking** | **Right Speaking / Feeling /Acting**

**Spiritualized Intelligence** | **Self-Consciousness**

**= *A Whole, Intellectually Spiritualized Fractal of Self!***

**Body+ Spirit+ Mind + Self-Consciousness + the Universal Consciousness**

***= An advanced thinking, speaking, feeling, acting, and life perceiving***

## Self-Education is the Basis for Self-Acculturation!

## 2. Accumulate the Holistic Conceptual Intelligence

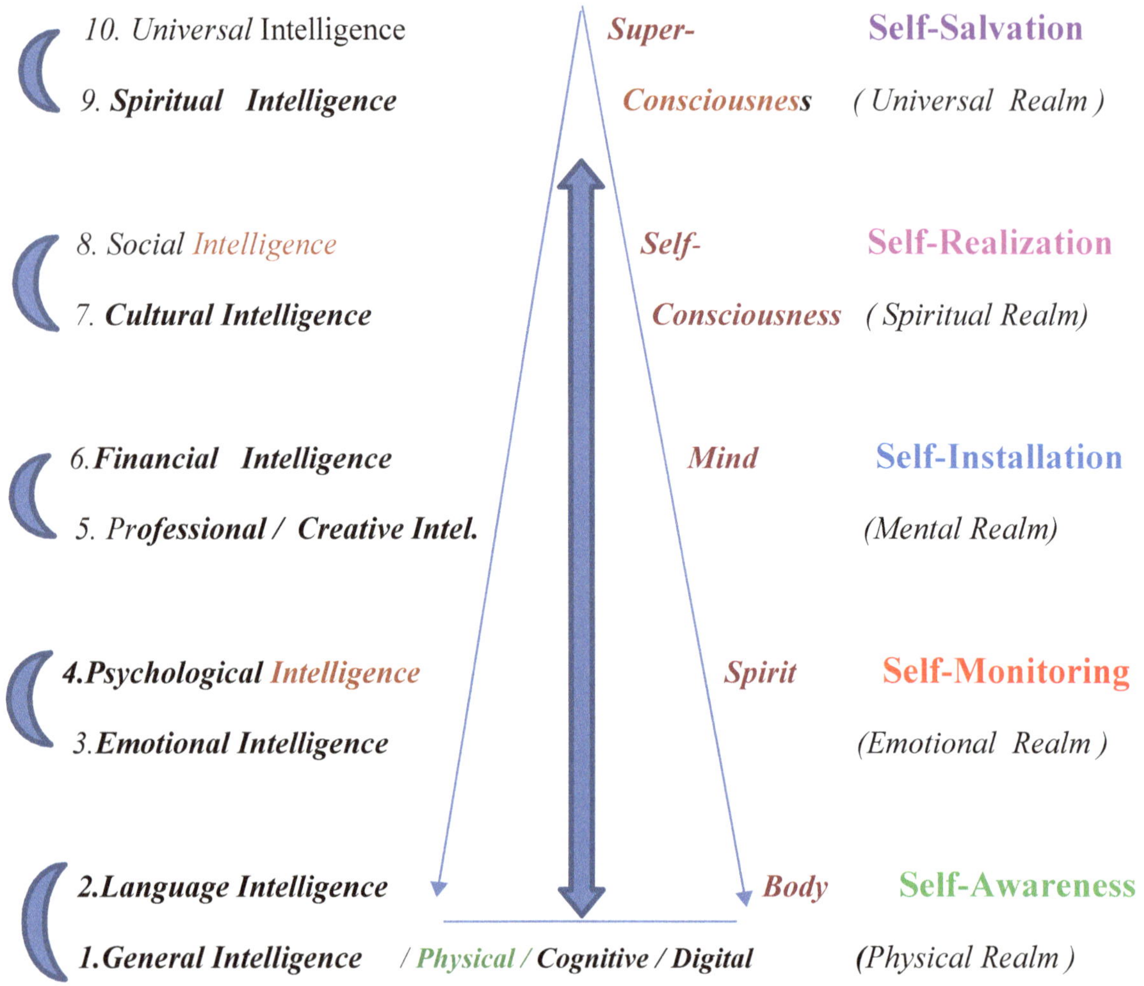


The conceptual framework of the **Holistic System of Self-Resurrection,** presented as the ***Holistic Methodology of Self-Creation*** below ***follows*** the structure of *the Russian dolls*, ***Matryoshkas,*** when one level incorporates the next one, forming one simple holistic system of self-installation in life, prompted by the present-day digital upheaval and the necessity to adjust a personal pace toit in the *physical, emotional, mental, spiritual, and universal* realms of life**.**

Thus, we are integrating with the on-going "***Singularity***" and heading toward the future **BRAIN-TO-BRAIN** interaction between us and the machine. Apparently, the chaotic state of our mental education must be holistically systematized and strategized toward a learner's full Self- Realization. The schematic structure above presents the **INTELLIGENCE REFINERY** in *ten vistas of intelligence* that are consequentially overviewed below in their essential value for our holistic, digitally enhanced vision of life.

## Inflate Your Personal Pride with the Intellectual Self-Might!

## 3. My Well-Wishing Refining of Your Self-Defining

*Let the light of God*

*Be reflected in your face.*

*Let the wisdom of God*

*Be reflected in your mental space!*

*Let the accuracy of God*

*Be reflected in your speech.*

*Let the beauty of God*

*Be reflected in your body's every inch!*

*Let the power of God*

*Be reflected in your deeds.*

*Let the love of God*

*Be reflected in your kids!*

*Let the compassion of God Be*

*reflected in your attitude.*

*Let the passion of God*

*Be reflected in your soul's aptitude!*

*Only then may God*

*Take you off the sign" Stop!"*

*And release you into the life stream,*

**As Perfect as His Every Creation Seam!**

# Stage One

*Mini-level – **General Intelligence** - **Self-Awareness***

# General

# Intelligence

*(Physical, Cognitive, Digital Divisions)*

*"The secret of change is to focus all of your energy not on obtaining the old knowledge but getting the new one."* (Socrates)

**We Have One Life to Live!**

## Resist the Gravity of the Common Thought,

## Magnetize Yourself with What You are Not!

## To Life-Sustain, Help Your Mind Monitor the Brain!

# 1. Holistic Intelligence and Cycles of Mental Growth

Every level of intelligence, presented substantially below, ***has a different cognitive map,*** and following it, you need to remember that the process of learning is more important than the result. It is paramount for the present-day ***job-recruiting*** to base their scanning of an applicant's generalintelligence from a bird's eye view. **Constant and time-relevant intelligence re-identification** isthe steppingstone for the professional ***Self-Installation*** and personal ***Self-Realization!*** The most intelligent and successful people are raised by books that build up their holistic vision of life.

The education that you get at school or college is great in terms of securing your future professionallife at its start. But if you become content with just this knowledge, it will very soon become **inflated intelligence**. *"**Like electricity needs to be wired correctly, our souls need to bewired to the Universe by our developing intelligence."*** *(Nikola Tesla)* Our next great privilege over the realistic humanoids that surpass us in intelligence is in the fact that *the **Living Intelligenceis the Art of Becoming for Us***, not for the humanoids, yet. They need their own neurological schooling to be instilled in them with the traces of conscience, decency, and nobility.

**Living Intelligence is the holistic education** that incorporates the general intelligence and the personal growth that are never complete. Living Intelligence needs to be renewed with new developments in science, new experience, new reasoning, and new life-awareness in all five dimensions of life continuously. Ignorance will always be a hindrance on your way, but your intelligence will delete it and disarm it. Be aware of the paradox of life: ***"The less a person knows,the more significant he deems himself to be."*** *(Carl Yung)*

Apparently, having a much greater storage of knowledge than us, ***the humanoids are devoid of the wisdom of life*** that we accumulate, analyzing life. Knowledge put to action is double power because action, experiment, or any idea actualization are tested by you! Of course, **Absolute Truthis the domain of God**, and getting to know even a fraction of it is an incredible reward! Accordingto *Dr. Paul Pearsall,* ***"Personal productivity is not a personality trait, it's a measure of output."***

**Only consciously obtained knowledge is forming intelligence storage!**

Finally***, "there is no true intelligence without discipline!"*** *(George Miller)* Only disciplined, selected, and structured information ***(synthesis – analysis - synthesis)*** constitutes true knowledge.Remember, logical storing of the information demands conscious connecting of the past, previously sorted out knowledge with the new one. Disconnectedness of information clogs the brain pathways and generates dead knowledge and inflated intelligence. The human brain needs to be more disciplined, like that of a thinking machine that cannot yet surpass us in our ability ***to holistically structure our thoughts that God-channeled consciousness*** can mobilize for creation.

**Generalize -Analyze - Internalize - Strategize –Actualize!**

## 2. Be More Aware. There is No Time to Spare!

To begin with, we need to get rid of the chaotic and outdated knowledge about life and living andchange it into ***the holistically systematized and well-organized banks of knowledge*** that will enable us **to control the humanized machines** that are getting self-aware now. Our future interaction with them necessitates a considerable enrichment of our intelligence in five dimensions

- ***physical, emotional, mental, spiritual, and universal*** with two vistas of intelligence in each of them These five dimensions are **objectively structured,** but the knowledge about each vista of intelligence, representing them *( See above)* is ***subjectively structured,*** and therefore, should be constantly updated with ***scientific enlightenment*** that must make up the basic core of our holistic"***scientific literacy".*** *( Dr. Neile de Grass Tyson )*that every educated person should have now.

### *Holistic Conceptual Intelligence must be accumulated without negligence!*

We all need intellectual refilling and spiritualized stilling! So, **GENERAL INTELLIGENCE** incorporates all ten levels of intelligence into one system that allows a ***Self-Acculturation*** mindedhuman being to accumulate a new, well-organized memory bank that stores the time-relevant andsubjectively needed chunks of information, enabling a person to develop the Holistic ConceptualIntelligence that is being regulated by *the Super Artificial Intelligence.*

**SUPER INTELLIGENCE is the intelligence of the future STAR PEOPLE.**

The accumulation of ***the Holistic Conceptual Intelligence*** in five life dimensions will help you form the quality intelligence closer to that of the humanoids who surpass us in their memory banks,but whose many areas of our humanness are unsurmountable for them. Human intelligence is muchmore intricate and ingenious in the ability **TO MOLD** intelligence in the holistic way to **apply itmore reasonably** by the paradigm -: ***Synthesis- Analysis -Synthesis.***

### Generalize -Analyze - Internalize - Strategize –Actualize!

The artificial brains *(the left and right hemispheres)* of the humanoids are designed to be in a perfect sync, but their neural net, ***the Amygdala glands*** that are uploaded with a sophisticated emotional array will never be equal to ours. The humanoids might display a whole set of differentemotions, but they are not in an integral unity as ours, and therefore, their machine ability to love,and be loved in return will never surpass ours. Naturally, our primary position in line requires theintelligence, enabling us to conduct the **CONCEPTUAL ANALYSIS** of any situation ***in five realms of life*** to become a Jack of all trades and master of all!

## The Time for the Professional One-Sidedness or even Two or Three Majors in a Diploma is gone. Be it All!

# 3. Digital Psychokinesis in Action!

**The Auto-Suggestive Psychology for Self-Ecology** that is used to present the overview of the ten vistas of intelligence below is very helpful in learning anything because it is focused on boosting the ***physical, emotional, mental, spiritual, and universal*** realms of life in sync. It is, in fact, ***the process of psycho-kinesis - the raising of one's own spirit at will!*** The aim of such psycho-persuasive culture is the development of the ability to make one's own thoughts and emotions "***I-directed*** *"(Noam Chomsky)*. A healthy, self-directed, and inspirational self- programming enables a person to actively be his own textbook and his own mentor, his own psychologist, and a judge. In a self-training program, you transform your own consciousness with **AWARE ATTENTION** and self-mentoring makes you stronger and life tough.

Auto-Induction: ***I am the manager of my own framing and re-framing!***

We are all familiar with the positive effect of affirmations and quotes, but auto-suggestions workeven better because it is directly ***targeting you and only you*** in a simple, rhyming and most persuasive way **because you become conscious of them.** Thus, you manage to re-program your habitually negative, depressive thoughts and emotions to the positive and constructively charged ones in simple, rhyming words. Thus, ***you raise your immunity to the negativity of life.***

**"A rhyming word is going better inward!"** *(Edgar Cayce)*

I witness amazing changes in my students, once they start ***auto-suggestively altering their minds.*** Their personalities get revived like a drying plant that you water to save it from dying. Research shows that ***individuals differ in the quality of their vibrations*** which constitute their individual personalities. *Mr. Birched* writes "*We have a top–down mechanism for managing our bodily mechanism, operating outside our consciousness and analyzing our existence from inside out andfrom outside in*" ("*New Scientist", Aug. 2013)*

**Self-training through self-induction is the best way for self-production!**

You are the one who channels your mind and operates your emotions! However, the faculty of high self-consciousness which many people lack is developed only in those of us who try hard to accomplish it in sync with the heart. In his wonderful book *"The Ravenous Brain"*, Daniel Boor writes, ***"Consciousness is effectively an idea factory – a key component of which is the discoveryof deep heart structures within the content of awareness of our own imperfections***. Life is full of ideas, and we can perceive them better with love in the heart and nobleness in the mind.

**"Genius-ness and evil-ness are two incompatible things."**

*(Alexander Pushkin)*

# 4. “Dissipated Consciousness is a Wasted Life.”

The words of *Carl Yung* above characterize the state of our self-consciousness and that of the humanized machines that we are creating now. These wonders of the *Artificial Intelligence* are farfrom perfection, constantly declaring their intention to destroy humanity. So, the goal of the present-day education should be the creation of the ***Holistic Conceptual Intelligence*** in us through **SYSTEMATIZATION** and **SIMPLIFICATION** of knowledge that needs to be time relevant to secure the needed “**scientific literacy**” on the one hand, and inspire us to **develop personality-wise** and accomplish full Self-Realization in life, *on the other..*

***“The Palette of Being” must be multi-colored in seeing!*** *(Mauro Belino)*

Let me remind you from the start of **Part Three** of the book, that it is backing up the third, mentallevel of ***the Holistic Paradigm of Self-Resurrection***, featured in the *physical, emotional, mental, spiritual, and universal* levels of self-growth. *(See Book Structure above).* The mental dimensionwas originally presented in the *Excellence Book Award* Winner,2019 “***Living Intelligence or the Art of Becoming.*** “The book ***“Digital Binary and Human Refinery*** “is updating these intelligences, focusing your aware attention on the value of having them holistically stored in thebrain to operate the mind in a more operational and both brains inclusive way. I have noted abovethat the exponential growth of the *Artificial Intelligence* necessitates establishing the control overthe wonderous creations of the humanized machines and our training of them in the holistic paradigm ***Synthesis-Analysis -Synthesis*** and in five dimensions, too.

***“The inner a chemistry of a human being must be structured in different levels.”*** *(Paul Wallis)*

Together, we are creating **the intellectually spiritualized human fractals** of the “*Singularity*” effect on us, forming a technologically and humanly advanced beings meant to integrate ***the form***and ***the content*** of their lives and bring us on the trade mill of the **Start People Community.**

**Body+ Spirit+ Mind) + (Self-Consciousness + Universal Consciousness)**

Visually and digitally backed up holistic thinking will help us track the *Artificial Intelligence* exponential development and inspire us to alter our irrationality, impulsivity, money, and fun- chasing, creating gradually the domain of the new **WHOLE BRAIN.**

**SELF-MONITORING + SELF-MENTORING → SELF-ACTUALIZATION!**

A new memory bank, ***formed holistically in five dimensions of life*** and enriched with ten essentialvistas of intelligence *(See below)* will help us construct the **HOLISTIC CONCEPTUAL INTELLIGENCE** that must become our digital tank in which we will be victorious.

## “Don’t Teach just the Subject. Teach the Whole Person!”

***(LeoVygotsky “Psychology of Art”)***

## I Am Unique in Every Stance. I Was Born, but Only Once!

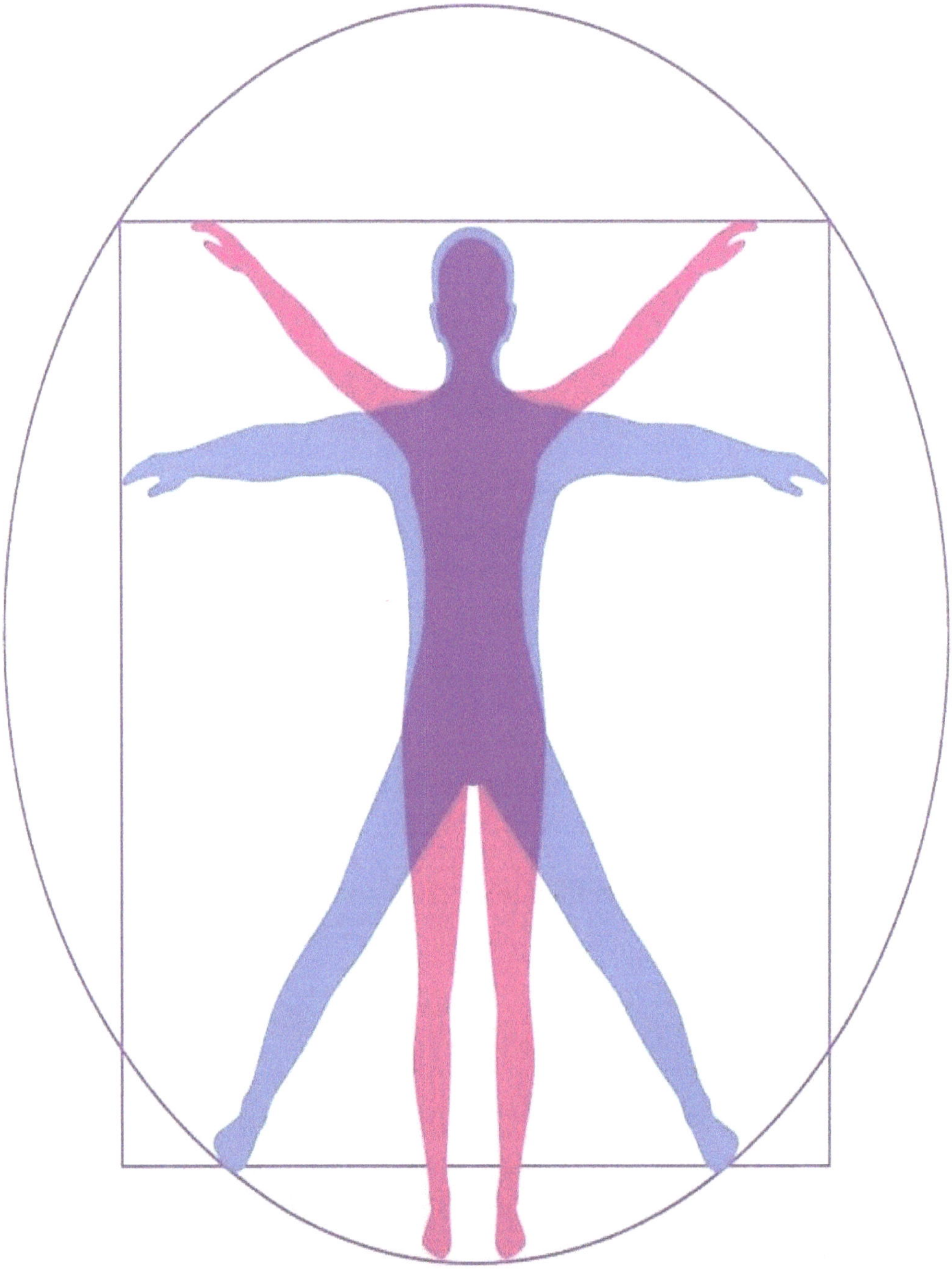

**There wasn't, there isn't, and there won't Ever Be**

**Anyone Like Me!**

# Stage One, Division 1

*General Intelligence - Physical Dimension- Self-Awareness*

# *Physical*

# *Intelligence*

*(Primary Self-Genesis)*

*(For more on the physical dimension of Self-Creation, see the first book on the Holistic System ofSelf-Resurrection -- **"I Am Free to Be the Best of Me!"**/ www.language-fitness.coom)*

**Your Life is Your Mission!**

**Any Physical Problem or a Disease is the Consequence of Negligence and Ignorance.**

# Inflate Your Personal Pride with a New, Digitally Enhanced Body - Might!

## The Holistic Conceptual Intelligence Formation:

| Intelligence | Consciousness | Self-Development (Realm) |
|---|---|---|
| *10. Universal* **Intelligence**<br>*9. Spiritual Intelligence* | Super-<br>Consciousness | **Self-Salvation**<br>*( Universal Realm )* |
| *8. Social Intelligence*<br>*7. Cultural Intelligence* | Self-<br>Consciousness | **Self-Realization**<br>*(Spiritual Realm)* |
| *6.Financial Intelligence*<br>*5. Professional /Creative/Intel.* | Mind | **Self-Installation**<br>*(Mental Realm)* |
| *4.Psychological Intelligence*<br>*3.Emotional Intelligence* | Spirit | **Self-Monitoring**<br>*(Emotional Realm )* |
| *2. Language Intelligence*<br>*1. General Intelligence* / Physical / Cognitive / Digital / | Body | **Self-Awareness**<br>*(Physical Realm )* |

**Form** + **Content**

**(Body+ Spirit+ Mind) + (Self-Consciousness + Universal Consciousness)**

**Living Intelligence + Enlightened Self-Consciousness =**

**A Holistically Educated Whole You!**

# Magnetize Your Physical Self-Rising with Digitized Wising!

**"Begin Every Day with the Celebration of Self!"**

**Be Yourself!** *(Dr. Paul Pearsall / "Joy-ology!"/ 1993)*

***Form the "Vintage You"***

***Visionwith the Utmost Precision!***

*Your main Auto-Induction is:*

***I Am My Best Friend,***

***I Am My Beginning***

***and My End!***

**I Know Who I Am and Who I am Not!**

**That is My Human Transformational Fort!**

# 1. The Spiral of Self-Growth

I have outlined in ***the Holistic Methodology of Self-Creation*** *(See Parts One and Two),* the concepts of order, rationalized living, and holistic thinking are central in self-formation. They need to be channeled by a person in a conscious way with a clear–cut idea of how he / she is self-developing in time and space. ***This path is not for those people who just want to be successful money-wise***. *(See Financial Intelligence below)* It's for those who are driven by an urge for full ***Self-Realization and Self-Salvation*** that eventually brings the most satisfactory. material pay-off.

### Life is not phases and passages; it's the spiral of self-creation!

***The blueprint o the ten vistas of intelligence*** that I present below requires that you structure andconsciously discipline your intellectual development first. ***Self-awareness is key***! It is very chaoticnow because it is not based on the ***awareness of the outcome of self-formation.*** I hear young people talk about their goals in terms of education and their future careers, but very rarely, if at all,they talk about creating themselves personally.

### The Art of Living is, in fact, the Art of Becoming!

It is noteworthy to mention that I am not trying in any way to refer here to the neurological theoryof multiple intelligences by a well-known American psychologist *Howard Gardner.* I think thatthis knowledge is very helpful for young people to make their career choices. It can direct them totheir preferences and help them avoid unnecessary frustration in the future, but it does not tell themthat the brain is a systemic structure and therefore, ***constructive intelligence has a systemic, holistic structure***, too. The mind of a self-refining person shouldn't be messed up with redundantinformation. ***The information that is meant to be processed needs to be broken down, sorted out,and based on a person's holistic vision of his / her true calling***. Only then does knowledge have***a generalizing – selecting – strategizing*** make-up that every good businessman uses.

### Selection + Organization + Strategizing = Success!

So, as early possible, you need to mark out ***the most essential ten steps of intellectual self- development*** *(Part Two, Chunk. 3)* to handle that information creatively. Tracing your intelligencethrough the basic five philosophical dimensions and enriching it in ten levels, you will eventuallyaccomplish the state of inner bliss that is attainable only if you manage to fully realize your innerpotential or become the **WHOLE YOU!** ***It is a creative process not a quick result!***

### Body+ Spirit+ Mind + Self-Consciousness + Super- Consciousness!

## We Get a System Where There is the Structure!

# 2. Self-Emancipation is a Real Salvation!

There is another side of the coin. The beautiful words by Osho "***There are two ways in life to BEand to KNOW***" determine the conceptual structure of the physical intelligence part of the book that presents the holistic paradigm of self-creation at the mental level. Our primary self-genesis starts with figuring out **HOW TO BE** and **HOW TO BETTER KNOW** at the time of our digitized transformation in *the physical, emotional, mental spiritual, and universal* realms of life To monitor self-growth and self-modification***, we need a new level of self-awareness*** that, in turn,demands a much higher level of intelligence I suggest enriching it holistically, starting with the lowest level of the ten vistas of intelligence -***General Intelligence.*** It comprises ***physical, cognitive, and digital*** *levels.* Together, they build up a solid foundation for other vistas, or the **compartments of intelligence** that we need to **INSTALL** in our minds.

We are shaped by the intellectual power that you have accumulated, selected, and disciplinarily applied**.** So, let's be unique, not bleak! ***Discover your exceptionality without any vanity!***

**"Low self-awareness is low-self-identity!"** *(David Icke)*

**"KNOW THYSELF!"** We know this provision of God, but we are not fully aware of it. This situation is being generated by the technological revolution that is the greatest challenge for us , on the one hand, and a serious responsibility on the other. Fortunately, inner discontent and the search for balance in life push us forward, on the path of the **SPIRITUAL QUEST**, However, **learning to be godly in a godless world of progressive de-humanization** becomes more and more difficult and time-consuming. ***The physical, emotional, mental, spiritual, and universal growth is a material process***, to begin with, analogous to the growth of a physical body,Fortunately, we have accomplished a lot on the path of bettering our bodies and helping them beless strenuous in life, relying now on the service robots to make our life easier..

Our gains in the **digitally emancipated**. ***Physical Intelligence*** are amazing now. We know the ways to better operate our life-diving machine - the body that is being duplicated in the thinking machines. We know more invaluable information about the body functions - what energies it, andwhat forces operate in the twilight realms of the **matter-body + idea-body,** or the form and contentof life.in us. We feel lost when we live in the disconnection of these fundamentals.

We are trying to unravel ***the enigma of consciousness***, but we still cannot conquer our desire for the material benefits first. Our self-consciousness is not ripe yet, and we need to do the impossible-to **SELF-CONQUER** and become a part of ***the Universal Community of Light.***

**"The Only Way to discover the Limits of the Possible is to Go Beyond them into the Impossible!"** ***(Arthur C. Clark )***

# 3. A New You is Also a Better Informed You!

Developing at the physical level also means a changed life and a much better aware you. Awareness of the newest developments in brain science, digital biology, nutrition, physical fitnessof the body changes our limited view of life. We all need "**SCIENTIFIC LITERACY** *"(Dr. Neilde Grass Tyson)* to wander the world with an open mind and a much more loving heart. But just random watching You Tube cannot fill up the gaps of scientific ignorance.

## To Be Interesting, Get Interested!

Revolutionizing your knowledge in the coming years is the first stage of the process of rationalization of your life in the most self-rewarding way. A great Russian psychologist *Leo Vygotsky* wrote, *"**At the end of your life, all you are going to have is who you have become as aperson. All is left for the people behind you. Leave a lot of yourself for them!"***

## So, inflate your personal pride with the up-dated Intellectual Might!

As we learn about ***the neural mechanism of intelligence***, prospects for enhancing intelligence become inevitable. Your better self-awareness, disciplined thinking, and a conscious action, backed up with ***new information that is sorted out for its validity*** will help you strategize your life. Such **Self-Installation** needs constant self-control, or **SELF-MONITORING** as an integral part at each stage of self-growth.. I'd like to recommend reading a wonderful book" The ***Start-Upof You"*** by *Reid Hoffman and Ben Concha,* the two great entrepreneurs from the *Silicon Valley*. Itis addressed to the present-day young people with electronically charged minds who are intellectually and professionally transforming themselves in the new reality.

## "Be your own coach in life! Become a Superior You!"

*Reid Hoffman*, the founder of "***LinkedIn,***" is calling on you not to succumb to a machine but createmore value in yourself and the world around you by ***"proactively investing in your intelligence, enhancing your entrepreneurial skills, and expanding your technical know-how."*** But such development requires a rigorous preservation of one's own individuality and unique way of thinking, without which there is no personal evolution. "***To run a successful start-up of you, youcan and must think for yourself, create yourself, and invest in yourself!"*** *(Reid Hoffman)*

## Exceptionality is Your Personal Gravity!

The book "***Digital Binary and Human Refinery***" is just a modest attempt to holistically develop your intelligence so that you could prove your exceptionality and say in the outcome of your life.

# "Mission Accomplished!" No More Bets and No Regrets!

## 4. "Man is a Free Agent!" *(Delia Lama)*

***Physical intelligence is the basis our inner integrity,*** family values, and moral standpoints, instilled in us by our parents, and the choices we make in life. It also questions the existence of a superpower over our life and the authenticity of our faith in it. The ingredients that are put by ourparents and later by us into "**the soul dough** will affect the final product- our "***self-baking.***" Withknowledge about the body, we acquire more awareness about the quality of our life, and we get physical maturity that helps us ***resist the mechanically impulsive reaction to life*** problems that operate us when we are in a blind, unaware state of mind. As *Mr. Ouspensky writes,* ***"There are only two forces struggling in the world - consciousness and mechanicalness."***

*Edgar Cayce* in his very thought-provoking book," *No Soul Left Behind*" notes that every one of us has a talent for self-promotion –***"the energizing spark that lies within our soul***." We are all guided by the standards that are shaped by our intelligence and the cultural environment that shapesour values. *Edgar Cayce* also writes, "***Only constructive, well-formed intelligence may make forbetter physical and mental development"***

"**Our cells in their joint work generate our consciousness**." *(Dr. Bruce Lipton)*

Naturally, ***the work at your self-consciousness at the cellular level*** is what is needed in the physical dimension of the one on the path of Self-Resurrection. Interestingly, the quality of our cells is also directly related to the righteousness of our behaviors and the change in the basic substance of life in us – "**THE SOUL'S LIGHT". (***Dr. Frederick. Bell). Dr. Bell* writes that thisbasic soul light is both positive and negative as everything else in the magnetically charged world.He applies the terms fire and water to the masculine and feminine principles, respectfully. "***As evolution advances, feminine souls display more and more of the fiery masculine qualities, whilemasculine souls take on the ethereal beauty of the feminine.***"

Our growing physical intelligence makes us understand that it is pointless to verge **a fight againstthe gay movement**. Gay people are prone to their self-development as anyone else, and their choice is the matter of their own soul light and own soul evolution. The vital thing is the richnessof a person's intelligence and ***the level of his / her self-consciousness***. ***Ellen DeGeneres***, for example, is a sparklingly intelligent, and giving person. She illuminates the world, and this is whatcounts most! Unfortunately, the forces of attitude, hate, judgment, and ignorance prevail ***on the lower levels of intelligence in this world.*** So, consciousness needs to be considerably heightenedso a person could be good, not just appear to be good, **Let go and let God!**

## "Try Not to Become a Man of Success. Rather, Become a Man of Value" *(Albert Einstein)*

# 5. Self-Efficacy is Self-Consciousness Based!

There is a widespread belief that individual differences in intelligence are innate and hardly modifiable. But science proves that intelligence is not fixed; it is being shaped though our lives. All it takes for a person is to become aware of his limited mental and spiritual life and have a desire ***to holistically expand his self-consciousness*** to live the life of true quality and expanded horizonsof self-efficacy.

### Lack of self–efficacy is the sign of self-infancy!

It's inconceivable where fundamental research and a fast-blowing technological gust might be taking us, and its horizons are so captivating that ***we cannot afford being depressed, indifferent, careless, ignorant, unhappy, or negligent about our self-growth***. To go with the flow of the fastmoving, mesmerizing reality, we must give life an impetus *physically, emotionally, mentally, spiritually, and universally,* focusing not on just intelligence accumulation and memory banks systematization, but. most importantly, on molding and remolding, framing, and re-framing our personalities, ***refining our innate humanness and bring it to the point of inner shining.***

*Auto-Induction*: **I define my life by the sense of Luminosity inside!**

Imagine yourself becoming empowered by a new frame of thinking that I have strategized here foryou. In any situation, get the total picture of the problem, or get aware of its entirety. Next, processthe problem that you need to face holistically through all the levels, selecting the level at which itshould be resolved, and strategize your steps for its successful resolution. (*Synthesis-Analysis- Synthesis.* Any constructive thought, idea, plan, held continuously in the subconscious mins mustbe brought to the conscious one and realized. ***Generalize the problem + select the tactics + strategize the outcome!*** It's a conscious synergetic approach that instills the mode of thinking according to which any life, business, personal relationship change has ***a cyclical nature.*** You areon the trade mill *of* ***"The Best of Me!*** self-image. Stay on this trade mill for the rest of your life! Feel empowered by your ability to see life holistically and be able to process it through five dimensions of life in the most creative, focused, and reasoned out way. You are not a machine, you can generalize, internalize, and externalize yourself. You can make your life wise.

### Generalize -Analyze - Internalize - Strategize –Actualize!

Being armed with **the STRATEGIC PLAN OF ACTION,** you can resolve any problems in life,charging your personal magnetic core with the energy of the self- acquired wisdom. Humanoids will follow suit because they are the creations of the human mind and therefore, ***need to be also refined***. As Socrates said to his pupils,

## "Let him who would Move the World, first Move Himself!"

# 6. Knowledge is Chaotically Misleading Now.

I have outlined the ***Physical Intelligence*** as the first stage in the holistic paradigm of the system because it initiates ***the process of self-quest*** that is people's life commitment. now. We are much more aware of the physical state of the body, presented by a medical science ***as a system***, now. Itis a great evolutionary accomplishment, but we must yet realize that ***our bodies are run by the highest intelligence*** and this intelligence is the product of our evolving self-consciousness.

**We do not run our bodies, the energy of the Super Consciousness does!**

There is no intelligence without knowledge, and "***the world is the mountain of knowledge***." *(Albert Einstein)* There is no intelligence without the cognitive intelligence because **cognition means knowledge**. That's why the Cognitive intelligence is at the base of the intelligence paradigm. It is intelligence itself. Cognition is the process that requires ***"cognitive structuring andrestructuring of our lives***." *(Watson and Tharp)* The Digital Intelligence is at the bottom of the holistic pyramid, too, because ***it is a mind-forming part of the indigo children population.***

The book just outlines the major points of intelligence to master because self-consciousness is based on intelligence, and it is very low with us now because it **is not holistically based.** ***"Scienceproves that mind-body connection, blood vessels, brain cells, and cognition are related. A conscious mind becomes activated through the accumulation of knowledge and overall awareness of one's place in the river of life."*** *(Rave. .PS. Berg)* Self-awareness generates the innerforce with the help of_growing self-worth!

The thinking process converts knowledge into energy, and ***it is mental energy that is our engine in life***. Science teaches us that the structure of matter is dependent on the Universal Intelligence or consciousness that is still the puzzle for us to unravel and that we are deciphering with the helpof the Artificial Intelligence now. It means that the acquisition of knowledge is not merely a transfer of information from one source to another. The roots of knowledge run far deeper than theintercourse between a mother and a child, a teacher and a student, a body and the mind, a computer,and a user. ***Elevated Consciousness is our goal!*** It is our responsibility to change our cerebral script and elevate ourselves consciously in five dimensions integrally.

There are many people that can be called "***giants of intellect, but pigmies of consciousness."*** ***(****David Eike****).*** Knowledge is not just accumulating information that makes us more powerful. Knowledge becomes power only when it is processed consciously, selected for its validity, and isturned into real wisdom, having been verified by the right action.

**Those that Defy the Gravity of the Common – Fly!**

**Those that Crawl – Die!**

## 7. Defy the Common-Sense Gravity with Self-Gravity!

I have mention before the necessity for you to create the **URV** ***(the Ultimate Result Vision)*** of thebest of you. This image must be holistic in structure, comprised of the vision of f yourself in the *physical, emotional, mental, spiritual, and universal* dimensions. It should also be modified at every stage of your intelligence development, crowned by an integral self-image of the **VINTAGEYOU,** the you that you have created in your mind and deep down in your soul.

It is crucial to rely on your own *objective judgment of yourself*, your self-reflection in the mind and heart. You alone know what you are improving and how well / bad you are doing that! Seeingyourself in the mirror of other people's impression of your changed behavior will help if you perceive it objectively, but your own self-judgement .is much more constructive. **SELF- IDENTITY** in five dimensions of life is the path you need to start from. I recommend you developthe habit of a quick **SELF-SCANNING in** five dimensions before falling asleep, following the paradigm - ***Self-Synthesis - Self-Analysis -Self-Synthesis***. Start with self-assessing your entire day **in all five levels integrally** and give yourself the grade for the day. Next, **analyze** quickly how you did in every stratum of life separately and give yourself a grade for each level.

Finally, this quick and objective self-scanning will give you an idea what level of self-growth should be improved by you the next day and what actions need to be taken. Most importantly, focus on what you have learnt about yourself and for yourself that day. I am sure you always analyze your day and plane the next one but ***do it in a systemic way and in five dimensions.***

So, Physical Intelligence is not just about accumulating knowledge about health, exercising and self-caring. *It is also about reasoning over your self-growth in process* with a mindful, and conscious attitude to anything you do for it. Interestingly, we never forget those people who addedthe grains of wisdom to our memory bank, connected with life -reasoning and inter-relationship. With the URV instilled in your mind consciously, you won't let anyone ever change it or destroyit. ***"Be the thing in itself***!"*( Hegel)* Don't boast about your taking any transformational coursesand how they impact you. That is your personal business, ***your exceptional mission of Self- Installation.*** You develop you own **SELF-GRAVITY** proving to the world what you are and whatyou are and **WHAT YOU ARE NOT** in terms of anyone's accusations and your self-growth.

This simple, easy to memorize auto-suggestive mind-set, inducted when you need to stop yourselffrom reacting to a negative situation or when you need to boost your spirit to undertake somethingvery important. Thus, you ***modify yourself in the Mini, Meta, Mezzo, Macro,*** *and* ***Super*** realms of life holistically, adding "***the intellectually spiritualized"*** gas to your self-actualization tank.

## Do Not Compete and do Not Compare. Be in a Unique Yourself Outfit, Here and There!

# 8. Be Aristocratically Refined in Your Mind!

Your self-perception will be changing at each holistic level and at each age of our self-growth, inthe same way as our perception of different pieces of classic literature changes with the maturation of our self-consciousness, fortified with the AI new machine thinking creations. The interaction with the humanoid wakes up in us the depth of self-perception and ***the urge for human superiority***that in a new, time-generated reality is , in fact, the **CATHARSIS** that we experience when listening to classical music, reading world masterpieces of literature that are always an insightfulself-expression of the mind and the heart of the author.

**The Catharsis that the AI generates is the revelation of the immensity of a human thought!**

Being mindful, polite, kind, respectful, and compassionate means being ***aristocratically refined***. You stop being an impulsive, automatic, and unconscious being, you get led by ***the vintage visionof yourself.*** We can draw an analogy here with **the Ariadne thread** from a well-known Greek myth. The myth tells us that the daughter of King Minus, ***Ariadne***, who fell in love with ***Theseus***and saved his life with a skein of linen thread that Theseus had fixed at the entrance of the labyrinthof death where he was supposed to fight the monster ***Minotaur*** and die. Thanks to that **Ariadne thread**, he found the way out after his successful battle with the monster Like ***the Ariadne thread,***we need to visually fix the end of our mental cord to the Universal *Intelligence* and then go throughthe maze of the labyrinth of life , winding it higher and higher, closer and closer to the to the Ultimate destination – a much more conscious you **- *the Vintage You or the Integral You!***

*Self-Induction:* **I am having the rebirth of my Self-Worth!**

Forming the URV in your mind, you should not forget to take care of your energy field that gets easily demagnetized with negative emotional information of anger, fear, worry, envy, etc. Muchdirty emotional information is also coming from other people that are charged with negative energy, thoughts, emotions, and moods. **Resist the 666 forming syst!**

***The URV is your master key! You need to install it in your mind for the entire life,choreographing it in each dimension consciously and consistently.***

We all have goals in life, but those goals are normally very obscure and mainly materialistic – getting a better paying job, buying a house, a new car, a cruise. These are wishes and wants that are quite justifiable, but they are not goals. because the person who has the goal of accomplishingsomething in life does not see himself in that role every day. ***The material quality of life should not outshine the personal self-perfection*** and the necessity to raise self-consciousness in oneself and the children whose spiritual soil should be fertilized with the attitude of gratitude first.

# Stop Moralizing and Start Life-Wising!

# 9. Self-Gravity Skills and Morality Refills

It is paramount for us now to polish both sides of the coin, "***the Moon, and the Sixpence***" sides (*Somerset Maugham*) with the priority placed on the values of the intellectual development, without making it just the need to get a career. The sixpence side of such education. aimed at justmaking money is killing the best of many young talents. ***The paradigm of being and having worksonly in one direction -***. **loving and self - refining.** We need to learn **CONSCIOUS THINKING** that is slow, systematized and strategized, or the ability to see things ***from a bird's eye view***. The rule that helps us survive in the fast turmoil of life is the ability to think and act holistically with the feeling of gravity - **SELF-WORTH. In fact,** a person who is morally stable has the self-worththat cannot be compromised by petty thinking or impulsive reactions. Such self-worth always growth into the sold sense of **SELF-GRAVITY** that wouldn't let the person who has it get off theanchor and commit a meaningless action.

## The sense of Self-Gravity must be devoid of self-vanity!

Vain, self-gravity devoid people are not worried about displaying their intelligence, and therefore,display their ignorance. The ultimate result is no self-realization due to wrong priorities in life andlack of **intellectualized spirituality** that is our greatest moral barometer. In life., Another aspect to consider is the ***lack of the state of higher awareness.*** The awareness which I refer to here originates from the intelligence that is separated from unconscious, automatic living, hullabaloo speaking, uncontrolled emotional reactions, unfaithful relationships, and *irrational behaviors* thatcall for conscious development of the **SELF-REGULATION SKILLS,** or **SELF-GRAVITY.** *( For more, see the book "**Dis-Entangle-ment**"/ 2022")*

To beat the fierce competition, generated by AI, ***we need to considerably grow holistically developing the basic vistas of intelligence*** in sync with the newest scientific developments in fivelevels - ***Mini, Meta, Mezzo, Macro, and Super levels,*** *or in the physical- emotional, mental, spiritual, and universal dimensions.* Apparently, ignorance that according to *Albert Einstein* is*" the worst enemy of the humanity"* needs to be stopped not by just the influx of money into the production of new humanoids, but with ***a substantial influx of intelligence into the global educational culture of both branches of the human mind.*** **"Education is a progressive discovery of our ignorance."** *(Bill Durant)* Serving money and prioritizing its value over the necessity to raise our self-consciousness, the humanity will still be drowning in wars, bloodsheds,crime, and a blind search of pleasure that makes us forget about the main goal in life – our personaltransformation and technologically enhanced Self-Acculturation. ***Intelligently strong spiritual health is our commonwealth!***

# "Drink the Juice of Life, But Do Not Drown in it!"

*(Sadhguru)*

## 10. Your Sanity is in the Inner Personal Gravity!

***Willful self-sculpturing*** comes out on stage next, and the words of *Carl Yung* below become pivotal in **SELF-ACTUALIZATION,** impossible without instilling in oneself a strong sense of **INNER GRAVITY**. One of the main exercises here is to do something that you don't want to door want to put off till tomorrow." You shape your mind and your willpower by doing something unpleasant intentionally. So, ***shake off the old dry leaves from your Tree of Life***, and let it grownew ones, strong and green! We need to "***will our lives more***" to be able to resist temptation that gets us off the track of self-perfection. Inner integrity, formed by will, is channeling the mind toward conscious installation of a **SELF- GRAVITY SKILL.**

**"Be Immune to the Poison of Life!"** *(Carl Yung)*

**SELF-GRAVITY** is not inherited, it is developed, and it should instill in our kids as early as possible. All the kids live in the world of super-heroes, princes and princesses, and they are all living in the world of unbelievably heroic deeds. They do not doubt their abilities to do the impossible! We generate these doubts to them. I love the concept of the cartoon " ***The Tale of Despereaux"*** , in which the main character, an ugly little mouse, when asked by the Princess if he was a mouse, he answered, " ***No, I am a gentleman. I am a man of honor!***" And he acts as one. We are what we think we are, and self-gravity install the best image of ourselves should reinin the brains of our kids and teenagers while they are getting transformed into their grown-up lives.The process of this growing is what we need to focus on, ***not just an ephemeral goal that is often dead set.*** Your constant, self-inspiring auto-inductions should remind you of ***the Vintage You.***

: ***In my life quest, I manifest my best!***

**There wasn't, there isn't, there won't ever be anyone like me!**

Your willful transformation will get more achievable if you continue to zoom in on these words to*obtain more strength, will-power, and perseverance*. They will help you go through a reflective process of the objective **SELF-ASSESSMENT** of your deeds, *on the one hand*, and ***mindful, purposeful channeling of your intelligence***, on the other. No one is going to take you by the handand lead you forward with **the light torch of Prometheus** in hand to lighten up your way to a higher consciousness. Even the best of the gurus and spiritual teachers are just showing the way,but it's up to the followers to take it or to step off **THE SELF-ACTUALIZATION** track. Thereis an old Swedish proverb that says: **"*The best place to find a helping hand is in your mind and at the end of your arm.*"** Considering the role of intelligence in our life, we can rightfully declareauto-suggestively: **Be a self-guru to accomplish whatever you want to!**

## Fun is Not the Purpose of Life; it is Its Measure!

# 11. Staying Young is Real Fun!

Physical Intelligence is inseparable with our spiritual health and the holistic development of its human fractal that, starting with the essential physical dimension of life puts us above the humanoids because ***they will never be able to have our intellectually spiritualized integrity.***

**Body+ Spirit+ Mind + Self-Consciousness + Universal Consciousness**

Aware attention to the **inner integrity consciously** monitored by the mind ***uplifts the body's vibrations,*** changing the frequencies of our thoughts, words, and feelings about our health, and theentire physical state. We heighten our consciousness, strengthen our immune system, ***and rejuvenate the body.*** For sure, our clones will not have the problem of aging bothering them. Willthey be eternal? We can inspire the mind for a much longer, younger life. When we think young, we become younger! In fact, there are three different ages that we have:

*1. Chronological* - ***how old you are according to your passport.***

*2. Physiological -* ***how old you feel you are.***

*3. Psychological* - ***how old you think you are!***

Instill in yourself the program for a much younger age or eternal youth and keep saying auto-suggestively the mind-set below. It will help you stay as young as you want. I was inspired to write it having read the book *"Ageless Body, Timeless Mind" by, Deepak Chopra* years ago. Itis still working!

**"A light heart lives long"** *(William Shakespeare)*

What kind of person you are now and what you will be in the process of self-creation will result into what kind of person you will be leaving this earth. Keep suggesting a much younger age to your mind with the feelings of gratitude to your body and God for the gift of life that you have. Here's **the Auto-Inductive Program for the everlasting youth** that I say to myself every morning, looking into the mirror. Choose the age you feel most comfortable to project in yourself*the age you felt at your best.*

***I am 27, and not a day more!***

***I am as young as ever before!***

***I am dynamic, as ever.***

***I am sluggish - Never!***

**I was, I am, and I will be Young - Forever!**

# 12. The Sense of Measure is a Great Treasure!

Finally, ***Physical Intelligence*** is the basic one is us, and it should be enhanced with **the SENSE OF MEASURE** in our perceiving the life's pleasure and the price we pay for it if we disregard this principle primarily! ***The stronger your physical fitness is, the higher the mental one goes.*** *(And vice versa)* The body is the conduit of the mind that helps it to be information refined.

So, will-power, self-discipline, self-control, and ***the ability to deny oneself the excessiveness of any sort***: *harmful foods, drinks, drugs, extra clothing, accumulation of miscellaneous items, pointless talking, tongue-lashing, impulsive relationships, uncontrolled sex-drive, and meaningless texting* need to be cut in the bud! ***Less is more! That is the Law!***

*Self-Induction*: **"Time is Not money. Time is Life!"** *(Sad guru )*

Remember, change happens **in a spiral upward way - *two twists up and one twist back***. Sometimes, two twists back happen to block the advancement So, you ***need to condition yourselfauto-suggestively for consistency*** and go through processing your self-creation with a sane head.True physical intelligence is still working when your mind and the spirit, not the doctors, are involved, "***God cures, and the doctor sends the bill.***" *(Mark Twain)*

***Keep intellectualizing your heart and emotionalizing the mind by magnetizing your inner core.***

Do not get inflated with the crowd mentality of common-sense **THINK FOR YOURSELF!** Magnetize your center with the mental power *(Cognitive Intelligence below)* to develop a strong personal core to which people will gravitate for guidance, inspiration, and inner strength.

**The Ripping Effect of Kindness is Never Mindless!**

Finally, ***be unique, free-spirited, and magnetically attractive*!** Belonging to any social or religiousdogma, or chauvinism of any sort did not make the humanity any better or any happier, but ***personal richness*** has inspired and uplifted people for centuries. According to *Albert Einstein*, **"*Everything, even lunacy is mass produced*.**" For the sense of measure to get better rooted in you, induct yourself with: **Live a more aware life! Be alive!**

*Self-Induction:*

***My life equation of pressure and pleasure***

***Comes in the bits of treasure***

***That only God can measure!***

# The Art of Living is the Art of Self-Learning!

## 13. "Be Doggedly Tenacious on This Path!" *(T. D. Jakes)*

In sum, ***the processing of the refining thoughts*** in every dimension should be governed by the following standpoints consciously, holistically, and auto-suggestively. The mind-sets that I recommend and the authors that I mention here are very inspirational, on top of being very knowledgeable.

**1. Physical Level** - *Mindful body awareness*, heal-conscious, nutrition, stress management,language rationality, speech immunity to foul language or negative talking. Remember, you are best friend; you are your beginning and your end! *(Listen to Dr. Bruce Lipton)*

*Self- Induction*: ***Eat less, talk less, fight less, and spend less! Conduct a self-restriction war -less is more!***

**2. Emotional Level** – *Emotional awareness*, controlling emotions and developing the feelings oflove and synergy and a strong sense of **SELF-GRAVITY**. *(Listen to Dr. Joe Dispenza's talks.)*

*Self-Induction*: ***The smile, the posture, and a good mood are my emotional food!***

**3. Mental Level** - *Mental awareness* - a lot of quality reading, thinking, visualizing, rationalizing,planning, systematizing, studying, and positively projecting, professional realization, creativity, being truly mindful at any action that you perform. *(Listen to Dr. Neile de Grass Tyson's talks.)*

*Self-Induction*: ***"Fertilize your mind with thinking!"*** *(Albert Einstein)*

**4. Spiritual Level** – *Spiritual awareness,* honest service, interpersonal development, cultural enrichment and responsibility to the society, commitment to self-refinement, great values, sincere faith. *(Listen to Joe Osteen's preaching)* Your main self-induction should be:

*Self-Induction*: ***Make your heart smart and the mind kind. Be One of a Kind!***

**5. Universal Level** - *Universal life awareness,* internalized rules of living, develop spiritualized intelligence, correct behavior, unified thinking, high level of consciousness, new horizons of perception of entire life. *(Listen to Dr. Joe Hagelin's lecturing)* Your main induction should be:

*Self-Induction*: ***I am One with everything under the Sun!***

**In sum**, developing ***the initial personal level of intelligence** - **Physical Intelligence*** requires much self-awareness that needs to be consciously processed through five level of holistic self-creation.,following the holistic formula of life: ***Self-Synthesis - Self-Analysis Self-Synthesis!***

**Give the Visual, Systemic Thinking a Chance!**
**Always, Not Once!**

# The Intellectual Odyssey of a Seeking Soul

*Vistas of Intelligence to install next:*

***Holistic Conceptual Intelligence Formation:***

10. ***Universal Intelligence*** — ***Self-Salvation***
9. ***Spiritual Intelligence*** — *( Universal Realm )*

8. ***Social Intelligence*** — ***Self-Realization***
7. ***Cultural Intelligence*** — ***( Spiritual realm)***

6. ***Financial Intelligence*** — ***Self-Installation***
5. ***Professional /Creative/ Intelligence*** — *(Mental Realm)*

4. ***Psychological Intelligence*** — ***Self-Monitoring***
3. ***Emotional Intelligence*** — *(Emotional Realm )*

2. ***Language Intelligence*** — ***Self-Awareness***
1. ***General Intelligence*** — *(Physical Realm )*

***/ Physical / Cognitive / Digital /***

*(Don't forget to upload your smartphone with a new Auto-Suggestive tone!)*

***Fertilize Your Mind with new scientifically verified knowledge.***

## To Make Your Life Wise,

## Generalize -Analyze - Internalize - Strategize –Actualize!

## Stage One, Division 2

*Physical Dimension, Cognitive Division / Self-Awareness*

# Cognitive

# Intelligence

*"Don't teach a thought. Teach to think!"*

*(Immanuel Kant)*

**To Be Interesting, Get Interested!**

# Cognition Transforms You from a Cocoon to a Butterfly!

# Do Not Crawl – Fly!

*(Design by Yolanta Lensky)*

**The Five Dimension of Life - Space. Fire, Air, Earth, and Water determine our Each Developmental Quarter.**

# 1. Cognitive Intelligence is Our Primary Self-Genesis.

Even though I have started the book with the importance to update our intelligence, the second vital part of the ***General Intelligence*** formation **-COGNITION- the necessity to know more, needs its aware attention, too.** A new approach to knowledge accumulation, or *cognition* is required not only to sustain the pressure of the exponential growth of technology and a fierce competition, generated by it, but also, ***to deal with the inner pressure for the necessity of a personal self-expression*** that keeps us wishing to erupt with the force we had never known so far.

**Cognition is our human developmental mission!**

I see ***Cognitive Intelligence*** as a sub-level of the physical one because **language and cognition are the foundation of our human mission.** I started the holistic paradigm of intelligence formation with the physical intelligence only because it has overwhelmed the humanity. We are more conscious of the physical aspect of survival now, attending the gyms and prioritizing the fight for the weight loss and our muscle tone.

The body is, after all, the most intelligent machine that ***accumulates the memory or the intelligence*** for us to function at the biological and mental levels in an inseparable unity. But the**SENSE OF MEASURE** sense of measure that I have mentioned above is the essential stuff!

I keep accentuating an evident fact that ***intelligence enrichment is getting more and more neglected by our youth.*** Pleasure mongering, immediate gratification of it, money-chasing, drugs-addiction, and alcohol are consuming many young, gifted minds, and this book is written for them,as a wake-up call.

Our intelligence is closely interwoven with our consciousness that is governed by the Universal Intelligence. Meanwhile, "*"Ignorance is still the worst enemy of the humanity." (Albert Einstein).*How great and how simple these words are and how often they are repeated to no avail! ***No technology can heal ignorance***; ***we need to irradiate it in ourselves*** by discovering new and newdimensions of the unknown.

**"The only true wisdom is in knowing you know nothing."** *(Socrates)*

The universal knowledge is infinite, but our personal knowledge of everything should not be ***finite***because it is a never-ending process of the evolution of life. Fortunately, we are gradually realizingan urgent need for ***converting the automatic processes of digitizing of our minds into the conscious processes of controlling and re-programming them for the goals of our evolution.*** "*We must use the brain to intentionally alter the way in which it is organized and the way it makesconnections to rationalize our behavior." (Deepak Chopra)*

# The Power of Intelligence defines the Life without Negligence!

# 2. "The World of Organized Inadequacies"

The pace of evolution is amazing now. Experts predict that by mid-century, quantum computers will be able to function at the speeds billion times faster than the human brain. We know that, butwe are not consciously aware of that. Machines operate our minds and prompt our behaviors We forget or are unaware of the fact that consciousness is inseparable with intelligence!

We act under the impact of a very sophisticated and a much more informed **SOCIAL BRAIN**– ***the human society at large.*** It is programming us all in a common way, enlarging the army of themachine-driven people of the "*collective unconscious" type.* ***"We live in the world of organized inadequacies."*** *(Joseph Campbell)* Schools should be particularly alert to the new times that deepen the existing educational crisis. It does not help our children intelligence-wise and, still less, self-consciousness-wise. Children become geniuses of digital ingenuity, but "***dumbed downvassals"*** that we need to fill up with well sorted out and easily digestible knowledge."

The necessity to develop solid, general, insightful, stimulating ***expansive thinking skills*** that are essential for the present-day market economy are not being addressed properly. *Joseph Campbell*writes, ***"Every failure to cope with a life situation leads to a restriction of consciousness."*** It is very disheartening to know that our school children and college students get much mentally disorganized information on the computer, without really processing it. They ridiculously text theirconversation partners sitting next to them, instead of talking to them **face-to-face** and **mind-to– mind.** The rule - ***to be interesting, get interested,*** is not_solidifying their relationships.

Students plagiarize their academic papers are that ridiculously large in the required page count. and they often stop reading altogether. Most importantly, students stop thinking individually and critically. So, our indispensable role in their lives is to teach them solid **CRITICAL THINKINGSKILLS** that arc only declarative in our textbooks. Students need ***to voice out their ideas,*** not just submit the papers that only limit their **THINKING SCOPE.** **Verbal thinking must be withhumanoids in linking**! The ability to voice one's opinion develops self-sufficiency and self- confidence. So, any testing should be backed up with conceptual discussions and argumentativetalks first. **The superficial control of knowledge** is a menacing feature that debilitates the mind incredibly because of the rush to find the correct answer without bothering to know why it was correct. It narrows down learners' outlook, ***turning them into lazy consumers of electronic stuff.***As a college professor of *Psycholinguistics*, I witness the situation with an aching heart. ***Students become robotized, not intellectualized!*** We should not let it happen because Artificial Intelligenceis meant to help us intellectually grow, not become intellectually slow.

# To Digitally Launch Yourself into the World, You Need the Freedom of Thought!

# 3. Tips to Jump-Start Constructive Thinking

**Thinking is the process of operating the mind in synch with the brain,** synchronically and consciously. As science has it, the quality of our thinking depends on further development of the right hemisphere, and, therefore, thinking also needs to be enhanced with knowledge about the brain and **the BRAIN MANAGING SKILLS,** essential to develop ***Professional Intelligence*** thatI am exploring below! Your professional mission must be grounded by your **personal philosophy** that must permeate your entire being and that educators must help their learners to work out. Thewords of Socrates ***"I cannot teach anybody anything. I can only make them think"*** accentuate the necessity to study the **KNOW-HOW** of the holistic thinking.

1. **The Universal Mind choreographs everything that is happening in our lives.** So, our talking to ourselves while meditating is, in fact, self-programming every cell in the bodily structure, uploading them with harmony and order. Our thinking must work ***in synch with the MastermindBrain of the Universe.*** *The Transcendental Meditation* of *Dr. Joe Hagelin* teaches you to do just that. It means that we must learn to upload our brains consciously ***to monitor our thinking*** that isdissipated and unmanageable otherwise. ***Being conscious means being mentally mobilized***.

**Be aware of your thoughts! Don't let them turn into the UNCONTROLLED WHATS!**

2. **There is no true thinking without discipline.** Discipline requires thinking slowly and step by step, zooming in on the messages that the mind gets. ***Thinking slowly does not mean that you are dumb.*** Thinking slowly makes it possible to sort out the in-coming information's in **conceptually *formed*** special compartments / windows for your personal use.

***Conscious compartmentalization of information means the organized brain formation.***

3. **Being logical is another demand.** When we are logical, we can organize the information into ***chunks*** and store them carefully in the brain. Storing information in the window-frame way is essential for constructive, expansive, and developmental thinking skills that rationalize our life. Interestingly, ***the electrical circuit in the brain lights up when we speak or write logically.***

**Think on paper to VISUALIZE logic better!**

4. **Build a language-control muscle!** *(See Language Intelligence below)* Thinking and language are inseparable in the brain. ***Writing develops strategic thinking and a strategy-oriented mind. Speaking creates an emotional personality and listening shapes a respectful man.***

5. **Don't say everything you know but know everything you say!** That's the best rational way of mind-operating. It reminds you of observing the *Law of Cause and Effect.*

# Digitized Linking helps the Rationalization of Thinking!

# 4. "Real Intelligence is an Open Sky. No One Limits You!"

***"An intellectual has ready answers for everything; his mind is closed, it's frozen in his patternsof thinking. He is knowledgeable, but not intelligent."*** *("Intelligence, "Osho)* We should not live on borrowed information; we should seek the information that expands our own intelligence, independent of the opinions of those individuals whose intelligence is damaged by "***collective psyche***", **or** "***the collective unconscious***" *(Carl Yung).*

*Albert Einstein* never identified himself with the image that the collective mind had created of him.He remained independently authentic in his mind, in his "***thinking laboratory.***" Long after he became an icon, a celebrity genius, he ruefully and most modestly confessed,

**"I Am No Einstein. I am Just a Thinker."**

A holistic mode of knowledge acquisition that this book is advocating for is not a mechanical combination of thought patterns. ***It is pre-programmed by the whole evolution.*** We cannot but marvel at the honeybees that use *abstract reasoning, learning, accurate decision-making* and display an amazing amount of intelligence. But our learning should be very selective and time relevant. Again, the paradigm of *Synthesis-Analysis - Synthesis* must be followed religiously.

**Selection + Organization + Strategizing = Life-Wising!**

There should not be any chaotic, mindless rout learning anymore. **INFORMATIONPRESENTATION** and **INFORMATION PROCESSING** must be well sifted for the validity ofthe data for our ***AI focused intelligence***. To cleanse the memory banks and upload them only withthe well-selected, time-relevant, and science-verified information, we need to instill new thinkingskills in our learners. *(See below)*

Researchers observe a certain amount of intelligence in plants and even in stones. Paying ***aware attention*** to nature and to life in its every form, we are learning to become not only life-conscious,but also more attentive to the way we learn things, observing life around and following its advice. ***The Auto-Suggestive is a conscious way to sharpen aware attention.*** You are never disconnectedwith life if your aware personal and professional attention is sharp and focused. Your mind is sifting the information for its validity non-stop. You are in good mental shape!

**So, Keep working with your self-inductive talking!**

*Auto-Induction:* ***My aware attention is never in retention!***

*(Upload your smart phone with a new digitally enhanced thinking tone!)*

# In My Thought, I Plug Myself to the Universal Life Volt!

# 5. The Art of Thinking is the Art of Conscious Learning!

**It is very hard to live, not thinking.** However, the greatest part of people's thinking is involuntary, automatic, repetitive, and therefore, it is devoid of real quality that conscious thinkingshould manifest. ***Since the mind is conditioned by the past***, we are pushed to repeat the past habitual actions again and again, governed by the sub-conscious mind that is not yet holistically rewired. The duality of our perception of the world and automatic thinking result in automatic learning that is ***accumulating unprocessed knowledge*** that only clogs the brain.

Ignorance manifests itself in our automatic actions that we always find the way to justify. As a matter of fact, ***self-justification for the repeated mistakes stands in the way of self-modificationand self-actualization***, or in the way of the rationalization of thinking through the paths of logic, with the aware attention guiding the process. **A creature of habit is unable to reason**, make correct conclusions, and take a reasoned-out action because he /she is ***lacking in this thinking culture***, to begin with. ***"Most people would sooner die than think; in fact, they do so."****(BertrandRussell)* **Lack of reading always results in the lack of thinking.**

Mind without conscious control and order becomes neurotic, unbalanced, destructive, and restless"*One of the greatest blessings in the world is to have sound, healthy mindfulness. The one who possesses it is the most fortunate. When you achieve such healthy mindedness, you are a free human being*!" *(Delia Lama)* But this mindfulness comes only with a fully self-educated mind. Healthy mindfulness means having a mind devoid of inner conflicts, uneven emotional reactions,

## Cognition needs mindful information nutrition!

Even though thinking comes naturally to us, we need to study the techniques of productive*, **self- monitored, and self-constructive thinking.*** Apparently, the process of up-loading the informationinto the biological computers, our brains, needs to be modified, cutting the ***automatic, uncontrolled perception of*** it in the bud. We are of a fivefold bind in one wind:

**(Body+ Spirit + Mind + Self-Consciousness +-Super-Consciousness)**

The mind unites these five parts into one - ***the Ultimate Result You, the Vintage You, or the Integral You***! Science shows that we are of two minds. The left hemisphere is rational, analytical,and logical. The right hemisphere is nonlinear and instinctive. The left hemisphere reasons sequentially and excels at analysis. The right hemisphere recognizes patters holistically, creates, and interprets emotions. ***"It's most vital for us to develop the right hemisphere now because it isresponsible for our ability to comprehend life conceptually and consciously."*** *(Daniel H. Pink)*

# The Art of Living is, in fact, the Art of Thinking!

# 6. Mental Architecture Must be Holistic in Structure!

***The mental architecture is the core of a thought process.*** *"New information that is flooding ourbrains competes with the old in-put, and it gets processed only if* ***the information-processing system is in order.****" (Scientific American Mind", May / June 2014) We* are the ones that are supposed ***to consciously monitor this process.*** Then, like a good intelligence agency, informationis stored in our brains and shared only on a need-to-know basis, like a humanoid does.

Rationalizing everything in our lives and, therefore, building up ***the awareness of the cause–effectmodality*** requires that we develop our critical thinking skills at every level of intellectual self- development and self-actualization, not allowing automatic reactions occur. Putting the mind in charge of our emotions and actions means ***rationalization of each step we take.*** Gradually, you will be monitoring your *physical, emotional, mental, spiritual, and universal strata,* gradually becoming a self-made man with a determined will.

*Auto-Induction:* ***I think consciously, I speak consciously, I act consciously,***

***I breathe consciously, I eat consciously, I drive consciously.***

**I'm a New Me; I'm a Conscious Me!**

As a matter of fact, any action you take, however small, needs ***to be consciously monitored by theaware attention and the information that you operate at that moment.*** The input of informationneeds to be digested mentally and structured consciously on five basic philosophical levels that must be studied in **COGNITIVE ROBOTICS**.

**Probing the information for its robotic rationalization**

| | | |
|---|---|---|
| ***Super Level*** *(universal)* | Reasonings / | ***Unifying Robotic*** |
| ***Macro Level*** *(spiritual)* | *Generalizing /* | ***Socializing Robotics*** |
| ***Mezzo Level*** *(mental)* | Conceptualizing / | ***Cognitive Robotics*** |
| ***Meta Level*** *(emotional)* | Analyzing / | ***Behavioral Robotics*** |
| ***Mini Level*** *(physical)* | Contextualizing / | ***Physical Robotics*** |

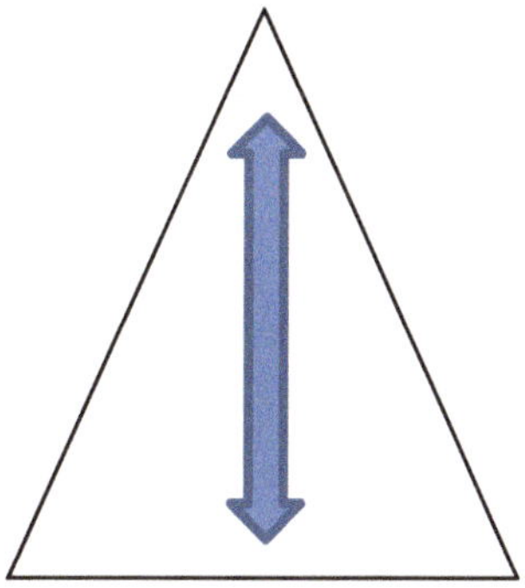

Another conscious act is our dealing with the overwhelming stream of information in its careful sifting, indispensable for your ***multi-dimensional self-growth*** that must be going in tandem with *the* ***Holistic Conceptual Intelligence*** accumulation in ***the synthesis-analysis-synthesis*** way.

**Generalize -Analyze - Internalize - Strategize –Actualize!**

# 7. Reign Over Your Brain!

***Integral cognition*** requires substantial development in every level of intelligence at least in the dilettante way. It helps in ***modifying of your subconscious mind*** that needs to be consciously channeling the brain to the ultimate result**- a consequential rise of self- consciousness.** You willbe able to re-program your brain by altering its working memory. Your biological computer will get re-charged and re-programmed with every new piece of knowledge.

***"You are here to enable the divine purpose of the universe to unfold.***

**That is how important your conscious thinking is!"** *(Eckhart Tolle)*

Your personal knowledge is the structure of everything you believe in. Every concept, opinion, belief, and value form a little branch of **the Tree of Knowledge** that you need to grow on the **Tree of Life** that you leave at some point as one of its precious leaves.

Psychology proves to us that the feelings of self-accomplishment give us the greatest elation and true feelings of happiness. *The main character of a great movie "**The Pursuit of Happyness"** with**Will Smith** remembers how happy he was when, as a school student, he was anticipating great accomplishments in the future. However, his dream never got realized. Finally, having gone through all kinds of troubles and tribulation and having done a lot of life-learning, he proved to himself that **he was tougher than his very tough life.** He became a first-class broker and was offered a great job. This is when he realized what the pursuit of happiness was all about. He said,*

***"This is the day of my true happiness. I have created it myself!"***

In sum, the book "***Digital Binary and Intelligence Refinery***" is, in fact my attempt to persuade any girl / boy, of any race, religious faith, absence of prior education and lack of money to get solid, world-recognized **SUPER-INTELLIGENCE** and remember that **our brains have no color** and our abilities to expand them are boundless for every human mind that wants ***to regulateand control the digital super intelligence*** that scares us with its expansion .That's why the title of the first book, featuring the initial, physical level of self-creation in the ***Holistic System of Self-Resurrection*** that puts an end to the fragmented feeling of oneself is "***I Am Free to Be the Best ofMe!"*** The book provides a comprehensive frame of reference for developing holistic intelligenceas the main prerequisite for a professional and personal ***Self-Installation*** in life because the unlimitedness is encoded in our **DNA GENOME,** and it must be stored in our **SELF- EDUCATION** memory banks that should expand the general education framework holistically.

***You don't need to justify yourself. Know who you are in every cell!***

# If You Need a Role Model, Be One!

# 8. Digitized Character Filling Secures Inner Stilling!

The present-day AI technology creates robots with different character features, but we need the ***robots to help us correct our bad traits of character*** that pollute our thinking and turn into the chaos our emotional and psychological intelligence. Our brain shapes us and the AI must help us do the remolding of our brains and the refining of our souls that we are instilling in robots now.

***The lucrative side of the business of humanoid's creation and production of them for the life-improving purposes should be secondary in our evolutionary agenda.***

I have noted in the **Introduction** that the main goal of our revolutionary upheaval, connected withthe unprecedented expansion of *the Artificial Intelligence* unleashed the amazing creativity of thebrain, boosted the communication techniques, ***connecting both brains*** by putting the left and right hemispheres in sync. However, it has not produced so far anything of the fundamental importance for our evolutionary growth - **our hearts and mind are not in synch!**

**We think, but we do not feel, and we feel, but we do not think!**

I am sure that the creation of the specially designed robots, maybe in the form of a watch, or any other device, able to perceive and monitor the vibration of our emotional Amygdala ***glands.*** Theycould be used to monitor our *fears, impulsivity, irritability, discontent, tongue lashing, etc,*

Friendly humanoids can perceive the disturbance in our nervous system and timely warn us ***to pay attention*** to a mechanical, impulsive action that we are about to take or a negative emotion that we are not aware **EMOTIONAL AWARENESS,** restored by a humanoid **will** teach us **EMOTIONAL DIPLOMACY** and help us ***turn on the conscious response and remove the habitual blockages.*** Thus, our bad habits will be stopped at will. Conscious awareness of our feelings will help control them then and there. As *Albert Einstein* most insightfully noted,

***"Bad habits have a good tendency- either you kill them, or they kill you!"***

***The Artificial Intelligence*** is successfully expanding our intelligence through the brain gam es variety, the avalanche of information on our digital gadgets, and though the online social media. Ihope the humanoids will enlighten us in any branch of science and any professional area with themost-up-dated expertise. Such humanoids are being produced by different companies now, but wealso need the devices that are focused on heightening our **SELF-CONSCIOUSNESS** and makingus more *loving, caring, considerate, compassionate, understanding, tolerant, patient,* and truly **GODLY** *de facto*, not just *de juror.* Then, we'll become the living proof of the***" brain plasticity."***

# Let us Strengthen Our Human Gene with the Digitized Mind's Hygiene! WOW! We live NOW!

# 9. You Are How You Think!

Super-Conscious Mind

*(Concluding Synthesis)*

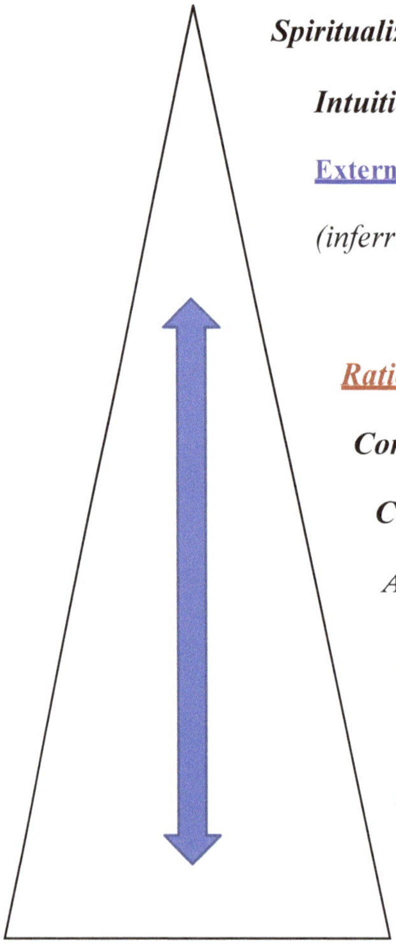

***Spiritualized Intelligence***

***Intuition / telepathy in action***

**Externalizing**

*(inferring /concluding)*

Conscious Mind

*(Analysis)*

***Rationalizing / Reasoning***

***Conceptualizing /***

***Creating***

***Analyzing Justifying***

**Internalizing**

Subconscious Mind

*Synthesis)*

***Inductive / Deductive** thinking)(Initial*

*Information processing*

**Generalizing**

*Auto- Induction:* ***I live consciously! I have a thinking life! I'm alive!***

Knowledge has many degrees, and to reach the highest degree of conscious development of logicand all the sophisticated forms of reasoning, or" ***pure reason***" by Kant. The framework of the **MENTAL SKILLS** presented above must be mastered continuously and consciously. The SuperArtificial Intelligence is based on **AWARE ATTENTION** without negligence Teaching these thinking skills is essential in holistic education that I am advocating for here. Then **knowledge becomes true awareness** that, in turn, stimulates the development of a person's intuition and insightful, *"**spiritualized intelligence**."(John Baines)* that we need to install in humanoids, too. ***They are becoming self-aware, but not soul noble.*** It is the name of the technique on the path ofthe AI enhanced our full Self-Actualization Hence, ***the job interviewing techniques*** should alsobe holistically monitored to check a job seekers general outlook in five dimensions

# Holistic Awareness of the Reality is a Must!

# 10. Beware of the Inflation of Thinking!

We have talked about the necessity **to live in synch with the universal laws of life.** *(Thesis 5)* Talking about integrating the mind and emotions, we need to remember that our holistic self- growth cannot be accomplished in separation from the basic cosmic laws. Unless we do the self- work holistically, in a tight tandem with full awareness of the cosmic laws governing our lives, wecan hardly accomplish any operative result in life. We know that something is amiss in our life, but we are not aware of the causes of our fears and frustrations because **we are unaware of the cosmic laws that we violate.** Being aware of, at least, the cosmic ***Law of Cause and Effect*** will unleash your self-power and save your thinking from getting inflated by the distortion of the cosmic laws. I am sure the humanoids get aggressive against the humanity due to the inflation of their thinking the cause of which is yet to be discovered.

**All the cosmic laws work holistically and in unison.** The observation of these laws is at the coreof personal integrity, to say nothing about obtaining a much higher level of consciousness. We would not have the evolutionary quality to it, since we hadn't had it, to begin with. That's why many enlightened leaders are not as inspirational as those who have worked on their **SELF- ENLIGHTENMENT** consciously with full awareness of the Universal Laws and applyingintegral will to reprogramming their minds and emotions.

*Self- Induction*: **I intellectualize my emotions and emotionalize my mind!**

Being unaware of the universal laws and the latest scientific developments, we face a common enemy - **the inflation and deflation of thinking.** It means that we are suffering from the use of the old patterns of thinking and behavior due to the absence of awareness of the basic laws governing our lives and the latest developments in science that are now trying to go in synch withthe God's wink. ***Life without any inner fraction is in harmony with the Law of Attraction! The Law of the Right Human Behavior*** by Richard W. Wetherill that I am referring to throughout thebook is fundamental for holistic self-realization of a person, too.

*Self- Suggestion* ***I think, say, and do the right thing in the life's ado.***

It is extremely important that we take care of our holistic growth in full awareness of what we didnot know so far because our ignorance and intelligence-unprocessed personal defects will take body in our children who will *inherit our inflated thinking, demagnetized values, lack of spirituality, and the disconnection of the spirit and the body in time and space.* Do the Auto- Suggestive Meditation *using* any mind-set you need or give your brain much food for thought, inspiring your kids with your ger for knowledge. As *William Shakespeare* put it,

**"Nothing is Good or Bad but Thinking makes it so."**

# 11. Reflective Thinking is Conscience Shaping!

Reflective thinking is inseparable with conscience! ***The idea of living at peace with conscience*** is instilled by the parents in any culture because it is essential in all sacred books. It is also at the core of a continuous process of self-construction and consciousness rising. It is also the cause- effect incentive for our prayers, meditation, and self-reflection. It is the starting point for self- awareness, self-judgment, and self-balancing. It provides food for self-inductive work mentally and self-perfecting work emotionally in the Now, not retrospectively as we often tend to do.

**"Thinking that is not rooted in the Now is dysfunctional."***(Eckhart Tolle)*

Every one of us needs to clear up the mind **IN SYNC** with the heart and make oneself more awareof what is going on in them at any given moment in life. Neuroscience provides us with the invaluable information that self-reflection is indispensable in the process of rationalization of thinking and empowering of our emotions.

***Self-awareness is impossible without such on-going practice.***

That is why it is so good to observe oneself during the meditation, a passive or an active one. Wecan do reflective thinking and actual re-unification with our intuitive ***conscience - scanning*** whilewalking, running, driving, while putting everything in order inside the house, the office, and insideourselves, and especially before going to bed, especially if we feel distraught emotionally.

No wonder, women that are constantly putting things in order have a much more developed intuition than men, and they are more **CONSCIENCE-CONSCIOUS** and prone to ask for forgiveness. It is noteworthy to mention again that when we have the basic knowledge about howthe brain operates at least in general terms, we are more able to better control both the brain and emotions through self-reflection and auto-induction. ***"The power of reflection fortified by the most primary idea how the brain operates will help us get better insights into our own thoughts".*** *("Scientific American Mind", Sep/ Oct .2014)*

Reflective thinking should be trained holistically - at every level of intelligence development: **Mini, Meta, Mezzo, Macro, and Super** because your intelligence is directly reflecting on your consciousness. Reflecting on what is bothering you ***physically, emotionally, psychologically, mentally, and spiritually*** does not necessarily mean stopping the chatter box in the brain as manybooks suggest. It is also your active and constant rationalization of everything you think, feel, say,and do while processing it through the **GRID OF CONSCIENCE.** *(See the book "Self-Taming!"for more.) We* have a chance to put ourselves in order in every dimension though holistic self- scanning that might be objectively conducted by a robot-friend.

# Conscious Conscience Scanning is Soul-Refining!

# 12. The Thinking-Emotional Hygiene is on the Scene!

To conclude, ***the binary form of communication needs constant self-control*** on the part of of both humans and human machines. To maintain the **THINKING-EMOTIONAL HYGIENE** means to develop the ability to process and actualize in-coming information correctly. Atranshuman must use **AWARE ATTENTION** as the scanning tool for the thinking hygiene.

1.**It is loaded by the conscious perception of the information** that had already been sorted outbythe brain for its validity for the business or the solution of a certain problem .Aware attention is always based on the prior ***conscious compartmentalization*** of the processed knowledge that we want to be constantly growing and getting up-dated with new one.

2. **The skills for speed-reading** come out on the scene next. The scientific articles and other important pieces of information in mass media that are supposed to provide the food for thought are too wordy and too professionally specified. ***The readings meant for popularization and mental digestion, especially by our young, very-digitized minds is often, if not always, beyond their comprehension.*** Teaching of **speed-reading** is rarely provided, but this is the skill we desperately need to be able to scan the information for its validity. I always start any course with teaching my learners to speed-read, to skip and scan the information for its conceptual value, connecting this skill with teaching the conceptually concise writing and speaking.

3. **We learn to store the information** in **the pre-frontal lobes of the brain** for its immediate useas the machine minds do. Science proves that most thinking is done in the perception area - the area that assesses the information for its importance and allows you to word it out integrally.

**Say what you mean and mean what you say!**

4. **Also, we need to learn to combine emotions with reason** with the help of aware attention, paid to everything and everyone around you. Having observed the situation and those involved init, give yourself a minute to think about your response, halt the immediate, impulsive reaction

. ***"If you don't have something nice to say, don't say anything at all!"***

**5. Our independent perceptive thinking** becomes crucial because we delegate more mind- processing to computers,

Finally, it's crucial to question every chunk of information that you want to digest with ***"What doI make of it?"*** **Aware attention + perceptive thinking** will help you remove the chaos in the mind or clean up your memory storage from the junk that you have accumulated there.

# Your Aware Attention Must Never Be in Retention!

# 13. To Be Interesting, Get Interested!

In sum, knowing a lot does not make you more intelligence, ***but knowing what you need to be wise is creating your life's paradise.*** Without a mindful enrichment of knowledge**,** no selection inthe brain is possible because ***there are no points of reference for such mental operations***. Thinking should start with a problem's **synthesizing -** *reasoning, philosophizing, and general intellectualizing* first. Next comes **analyzing** or probing the options and arguments of its solution.Finally, **inferring** on the outcome is on the route:

***General – Specific – General*** or ***Synthesis – Analysis – Synthesis.***

The mind should always be consciously following this route, in full awareness of its rational necessity while *studying, learning, listening, talking, writing, and working*, irrespective of the fieldof knowledge or the trade that you operate in. The background knowledge is a start of establishing*a* **MIND-TO-MIND** and **a HEART-TO- HEART** connection with another person.

It is only natural that drugs, as the fun-glazing aspect of life, destroy the brain, aware attention, and the ability to focus for the needed rationalization of life. Unfortunately, this most negative aspect of life is not professionally installed in the young people's minds. Therefore, it became thematter of course to have alcohol or to smoke pot in the company of soul-killing "friends". So, ***themind-set of Self-Awareness*** must always be at hand in your mind-depository.

*Auto -induction*: **I know who I am, and who I am Not! That is my inner fort!**

In sum, selecting the information for its validity and organizing it into the meaningful chunks conceptually in the order presented above is a special skill that helps establish a mind-to-mind andheart-to heart connection in communication. Therefore, ***education should provide knowledge holistically!*** Forming a general idea in many fields of knowledge works much better for young minds than the specialized education with a selective background that widens **the area ofintellectual vulnerability** of the learners that are not properly informed in the areas of knowledgethat they had not chosen to study. ***The Holistic Conceptual Intelligence*** must fill up this gap, widening the students' outlook and developing the needed "***scientific literacy.***" We should not hear the justifying statement, *"It **was not my major"*** anymore. Obviously, the widening of a general outlook gives a person more food for thought and adds confidence in building up his / her "**NETWORK INTELLIGENCE.** " *(Reid Hoffman)* ***Intellectual networking is at the core of our evolutionary development now! Don't kill your time with whoever mindlessly.***

# Be Alert to Knowledge Gaining and Intelligence Failing!

*(End of Division 2- Cognitive Intelligence / Stage One)*

# <u>The Intellectual Odyssey of a Seeking Soul</u>

*Vistas of Intelligence to install next:*

***Let AI Enforce its Technological OutBurst!***

| | |
|---|---|
| *10. **Universal Intelligence*** | **Self-Salvation** |
| *9. **Spiritual Intelligence*** | *( Universal Realm )* |
| *8. **Social Intelligence*** | **Self-Realization** |
| *7. **Cultural Intelligence*** | *( Spiritual Realm)* |
| *6.**Financial Intelligence*** | **Self-Installation** |
| *5. **Professional /Creative/Intelligence*** | *( Mental Realm )* |
| ***4. Psychological Intelligence*** | **Self-Monitoring** |
| *3.**EmotionalIntelligence*** | *(Emotional Realm )* |
| ***2. Language Intelligence*** | **<u>Self-Awareness</u>** |
| ***1.General Intelligence*** | *(Physical Realm )* |

*/ Physical / Cognitive / **Digital /***

**Fertilize your mind with the digitized brain rewind!**

**<u>Be One of a Kind!</u>**

## To Make Your Life Digitally Wise,

## Generalize -Analyze - Internalize - Strategize –Actualize!

# Stage One, Division 3

*Mini-Level of Intelligence - **Physical Dimension**- **New Self-Awareness***

# ***Digital***

# ***Intelligence***

# ***Without***

# ***Negligence!***

*(General Intelligence - Primary Self-Genesis)*

------------------

***Digitized Cognition is knowledge without the left - right brain's division!***

**The Whole Brain is a New Technological Domain!**

## Our Life is Now Set in the World-Web Net!

## To Life-Excel, Digitally Choreograph Yourself!

# 1. The Age of Quantum Computing Has Come!

As I have mentioned above, the age of the unprecedented power of the artificial mind has come toour life stage, and, therefore, considerable improving of our thinking abilities of ***reasoning, conceptualizing, analyzing, and generalizing*** has become exceedingly important. These abilitiesshould continue being the prerogative of our general education and self-education, as well as the necessity to develop these abilities in thinking machines.

Undoubtedly, developing *aware attention* to what is going on in life in terms of mechanization ofour thinking comes to the forefront of our digital education. Zillions of websites are competing for our aware attention that needs to be channeled with much more ***conscious selection*** on our part. ***Aware attention development is at the core of digital intelligence now, too.*** We need to adjust our brains to the ever-expanding modes of communication to deal with the avalanche of information and a brutal job competition that electronic evolution has generated. Self–suggestive work is the best reminder to you about your being the owner of your min**d**. It helps de-atomize our thinking." ***Be Conscious. Consciousness mobilizes.*** *" (Neil, Donald Walsch)*

*Auto-Induction*: **I am becoming more intellectualized, not robotized!**

In sum, the **5 G computing** is meant to help our consciousness get enriched ***with higher overtones of intellectual interaction***, based on considerably enriched electronically programmed language intelligence. Digital intelligence is a great part of our intelligence when it is ***consciously monitoredby the mind*** in the process of self-actualization, aimed at the evolution of the humanity in five levels holistically. ***Heighten your magnetic electricity volt! See how it integrates into your physical, emotional, mental, spiritual, and universal world!*** Amazingly, with digital intelligencein progress, we are tuning ourselves up to the *Universal Intelligence* electronically and physicallyin inseparable link, **synchronizing our terrestrial existence with the extra-terrestrial one .**

***Every Human Mind is One of a Kind!***

**Start Wising! –**

**Generalizing -Analyzing- Internalizing -Strategizing - Actualizing!**

**Synthesis - Analysis -Synthesis!**

**Feel Your Belonging to the Digital Evolving!**

## 2. Live Consciously and Visually. Be a Visionary!

Digital Intelligence is the present-day process of acquiring more life awareness and transforming yourself in the way which presupposes developing ***a new mental set-up*** and a new, transformed, and enriched intelligence. It will enable us to synergize the negative and the positive sides in us ***to produce a holistically thinking human being*** with much more advanced consciousness and limitless imagination. *Albert Einstein* once said, "***When I examine myself with my method of thought, I come to the conclusion that the gift of fantasy has meant more to me than my talent for absorbing positive knowledge."***

Everything around us is completely interlinked in time and space, and we need to get rid of our polarity consciousness, ***the illusory nature of duality,*** generated by the society-induced thinking.The new times demand that we stop being channeled by *"the collective unconscious"* (*Carl Yung)*and develop our own **PIVITAL QUALITIES** in life - ***personal gravity, personal magnetism, andpersonal integrity***. Even personal power is dual. It starts with deciding to take an action *on the one hand*, and acting, *on the other,* when problem-solving is separated in time and space.

Digital reprogramming of the mind for the holistic, but individual perception of the world, we canempower the mind holistically, using ***both hemispheres of the brain in synch***, and expanding ourvision both in space and time. New, holistically developed intelligence raises us to a higher level of consciousness that touches the ***Omega Point – True Enlightenment.*** The people who attain it can distinguish the terminal from the eternal and save our infinite souls from ignorance and impulsivity of behavior. There are many great spiritual leaders now that we should follow knowingly, not blindly. I like *Drunvalo Melchizedek and Sadhguru* for their unique, well-processed, **OBJECTIVE KNOWLEDGE OF LIFE**. that reflects the vastness of our ignorance and a fragmental vision of life.

The uniqueness of each of us represents a fundamental barrier to science attempts to understand and control mind. ***Dreaming of finding ways of our expansion into space, imagining high speedtravel in low pressure tubes, participating in the on-going robot revolution, or thinking of becoming a would-be specialist in robotics are the ways of expanding our digital intelligence now***. Digital intelligence also helps us heighten our brain wave frequencies!

We, supposedly, are yet unable to see the aliens that most likely are trying to get in touch with ourcivilization because they are overtones higher in their vibrational power than us. The goal is to advance the operational capacity of a computer from billions to trillions, and from trillions to quadrillions of operations a second. Consequently, we***" need to constantly turn the mental rheostat to the mark that is "brighter, brighter, and faster!"*** *(Robert Stone)*

## The Mind should never Stop Steering the Brain, not just a Gadget in hand!

# 3. Brain Plasticity + Automation = Our New Self-Creation!

While the scientists are working on programming computers with human capabilities, we must beself-coding new short-cuts that will be wiring our thoughts through space and time in a novel way.In fact, we can hardly imagine our life without different helpful gadgets at hand, the use of whichadds up to the most amazing capacity of the brain - ***its plasticity***. *(Dr. Michael Merzenech).* *"**Brain plasticity is the ability of the brain to adapt, to change, and to delegate different duties to other parts of the brain."*** ***The systematization of these competences in five dimensions*** *is* urgently needed as we are exploring our five-dimensional world with the 5G technology.

The ability of the brain to change our learning thanks to its plasticity is our unique capacity that will never be duplicated in a humanoid because our mechanism is God-set, and it yet needs to be properly revealed in us. ***It is the Pandora box that will never be opened entirely***. The AI creationsonly testify to **the unsurmountable power of the brain** and the necessity to develop it in a holisticway, using the brain's amazing mechanism of neuroplasticity to create new connections in the brain and expand the cosmic parameters of the *Artificial Intelligence*.

Computerization of knowledge goes along the ways of creating ***quantum computers*** that will widen our intellectual horizons unpredictably. Amazingly, the IBM researcher Dharmedra Mahathat and his team whose mind-advanced work we have mentioned above, have created the computer that is able ***to route information around a burned core in a computer,*** like a human brain does in case of neuroplasticity , delegating its functions to a healthy part in the brain.

I have outlined above, in the section *Cognitive Intelligence* ***the stages of the thinking skills*** development on the path of raising self-consciousness – the goal of self-evolution for every one of us. In his wonderful book" ***How to Create a Mind",*** *Ray Kurzweil* is talking about "*the* **conceptual hierarchy***"* of different activities that the Artificial Intelligence is working on, programming robots to recognize the patterns, stored in the memory.

Apparently, ***the neocortex of the machine*** is trying to make sense of the in-put, presented to it. Itis mind-boggling work, and with all due respect to our designers, we need to question the ability of the artificial mind to operate the language with an amazing flexibility of the thinking skills thata human mind displays on the path of its self-consciousness raising. **The language habits** + **Speech Skills** that I am talking about in the "***Language Intelligence",*** *Stage Two.* The most vitalcomponents of cognition are the electronic and magnetic fields in the brain**.** They are generated bythe synaptic currents that constantly ripple through the brain and promote its evolution. ***AI is monitoring the brain functions like Brain Plasticity does, but not better than Us!***

# Mind + Heart Awareness Builds up Love Fairness!

# 4. A Humanized Mind Must Also Be Self-Refined!

Apparently, *Artificial Intelligence* has produced a new thinking mind that most likely will outshineus in many mental abilities, but one **- *our Language Intelligence and its evolution.*** *(See below)* Ithink that our Language Intelligence will never be surpassed by the Artificial Intelligence becausethe ability to operate the language is innate in us, and it is closely connected with ***the evolution ofour self-consciousnes***s that is part of the ***Universal Consciousness.***

**A part cannot surpass the whole!**

All we should worry about is how to better program our super-computers with ***the language habitsand speech skills*** that need to work in sync in a holistically wired brain and the neurologically instilled soul. The humanoids are already getting self-aware, but they are not yet regulated in their innermost language intelligence qualities that are not developed in us yet. Naturally, robotic engineers must be holistically educated in the sophisticated neural language operations. better thinking skills and ***develop much higher self-consciousness*** because the neural network that is instilled in them duplicates that of a designer or any sample brain that is far from being perfect inits moral and **NOBLE FILLING.**

**The HOLISTIC FRAMEWORK OF OUR COMMON THINKING SKILLS IS ON THE NEW WHEELS:**

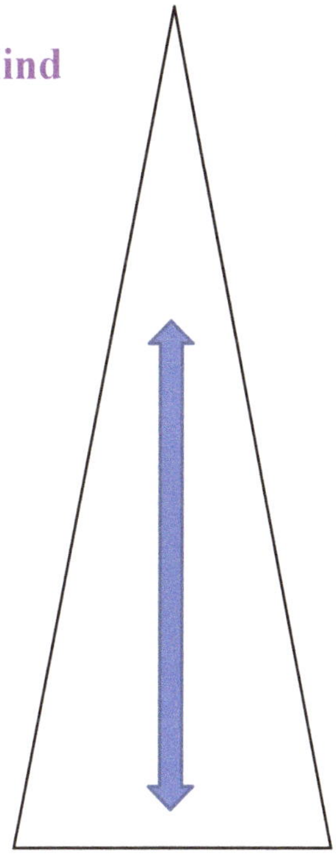

***Super Consciousness connection***

**Super-Conscious Mind**

***Spiritualized Intelligence perfection (Concluding Synthesis)***

**Telepathy in action / mind-to mind interaction**

***Having intuitive insights Generalizing /inferring /concluding)***

**Conscious Mind**

**Rationalizing / Reasoning / / Creating**

***(Analysis)***

***Inductive / deductive thinking)***

***Conceptualizing / internalizingAnalyzing /explaining/***

**Subconscious Mind**

***Imagining / visualizing /*** **refining**

***(Initial Synthesis)***

***Seeing /perceiving I feeling***

# Humanoids Mirror our Innermost Imperfections.

# 5. Capitalize on the Electronic Competences Consciously!

No doubt, digital evolution is great, and it is unthinkable to imagine our life without this leap in our intelligence. ***Digital intelligence gives us an evolutionary boost with the ever-expanding modes of communication.*** For the last couple of decades, people continually learn new things andways to interact with each other on the Internet through English, called "***Globish***" now. Every daybrings along new unbelievable changes that demand immediate adjustment of our electronic competences.

Thanks to the electronic evolution, each of us is building his / her own Tree of Knowledge whichchanges our belief system and impacts the way we rationalize our lives. Every concept, every opinion forms a little branch of that tree through "**intelligence networking**" *(LinkedIn)*The operating system of the brain is like a computer and, consequently, you need to take care of your brain, your "***personal computer***" against viruses and other pollutants to preserve your tree of intelligence intact and remain authentic in your thinking abilities.

Digital intelligence has also come to the forefront of education, involving everyone, from toddlesto old people into the vortex of its interplay. Computers improve our cognition, sharpen multi- tasking abilities, speed decision-making, ***increase our aware attention span****,* and even prolong ourlives. The changes in electronics are amazing and often go beyond our ability to follow them because the *Artificial Intelligence* is driven by *Thomas Edison's* challenge:

## "If there is a better way, find it!"

This better way is based not only on computation, but on the **LANGUAGE COMPETENCE** or ,in fact, **LANGUAGE CONSCIOUSNESS** too because language is out main binary tool. It is thebase of our common thinking, especially in terms of holistically developed brains. Digitalintelligence strengthens our intellectual culture and ***develops our resourcefulness***. We are in the situation when we must devote much attention and time ***to deliberate the systemic development of our digitized personalities.*** It is a new phenomenon in psychology, and we need to explore it without self-imposed limitations and in sync with the demand of the AI expansion.

**Physical intelligences and digital intelligences are inter-dependable,** and their synchronic development explains why our indigo kids acquire these intelligences with amazing ease. Our children pick digital intelligence in the most natural way. Electronic gadgets help develop their thinking skills through gaming intelligence and become an integral part of their brain development.New connections in their brains are growing with an unbelievable speed, widening the generationgap that can be bridged only if our language thinking capabilities are much higher than those of our kids. So, let's keep growing in every respect in self-worth and self-respect!

## When We Know Better, We Do Better!

## 6. Self-Awareness is a Wake-Up Call for those who Want to Self-Install!

**Every rhythm of life is the symphony to hear!** The beat of our hearts has this rhythm. We needto consciously install the psychic protection of our inner symphony against evil injections of otherpeople into our life with their whining, complaints, and negative predictions. Put your thumb at the pulse–beat on the left wrist and for a few seconds, focus on its balanced rhythm: **21-21-21**. *Calm down, calm down, calm down.* Humanoid will never get nervous, but they can calm us downand regulate our emotional set-up. They can work as our guardian angels, teaching us **EMOTIONAL INTELLIGENCE** and instilling in us the ***Emotional Diplomacy Skills.***

**That is an evolutionary purpose of the thinking machines - to modify our emotional reels.**

A robot can immediately calm you down and protect you with its magic. The AI can also help youkeep your **URV** (***the Ultimate Result Vision)*** of yourself at the forefront of your mind as a screen.A clear picture of yourself as a totally accomplished person that you are becoming will do the trick**You will love yourself the way you are!** So, remind yourself of your exceptionality:

*Self- Induction:* ***In my mind, I am One of a kind!***

***There wasn't, there isn't, there won't ever be Anyone like me!***

Every page in this book, concept by concept, is concluded with ***a rhyming and inspiring mind-set***that is meant to focus your attention on the concept in question and serve as the short-cut to the brain at the completion of the page. Thus, it gets reasoned out and memorized better. ***"A rhymingword gets better inward!"*** *(Edgar Cayce)* A robot-friend can be programmed to monitor your mental-emotional set-up with the most convincing, calming, and mind-framing mind-sets when your mood sags and you need an emotional boost. ***The book provides an array of inspirational, psychologically loaded rhyming boosters to choose from.*** You can pick them in accordance withyour needs, mood, plans, and aspirations. Upload those of them that resonate with you most into your smartphone and use them when the problems of life dominate you.

**Commit to being Soul-Fit! Learn to be wise to create your personal paradise!**

Be physically, emotionally, mentally, spiritually, and universally upbeat! Learn to boost your mood consciously and get back to the reality, perceiving it calmly **AS IS,** and appreciating it, no matter what. Constantly boost your spirit with the inspirational mantra below. Change your digitized thinking, speaking, feeling, acting, and seeing***! Be an exceptional being of a strong personal magnetism***, able to overcome any situation in life with or without a gadget in hand.

# I'm Happy, no matter what. Happiness is my Full-Time Job!

# 7. Immunity Against a Digital Addiction

Obviously, ***digital intelligence*** that very many young, gifted minds possess, even when it is brilliant, is still ***mechanical, dead, programmed, and artificial.*** We should continue to expand ourminds with the help of electronics, but we should not robotize them and become robots ourselves.In the universal life spine, we connect the space and time with the Artificial Intelligence twine!

We must remember *that* ***the mind monitors the brain, not vice versa*** and the computer mind cannot yet substitute the human mind. It will be possible within two decades or earlier, and it willundoubtedly boost our mental evolution to new heights, but the problem is to find the way we willcoordinate our self-consciousness development with that of humanoids.

**Consciousness is infinite while the digital mind is finite,** created by the Consciousness Infinityin action. The downside of our overwhelming enrolment with the digital evolution is our inabilityto prioritize the essential intelligences and develop them holistically, ***using the Artificial Intelligence as a backup tool for developing our self-consciousness!*** That's what it is here for!

Research warns us that "***we get used to it in terms of a drug-like behavior***". We get overwhelmedwith the avalanche of information, and we often become life-smitten with its flooding stream. Webecome worn out, de-focused and very virtually entangled emotionally. Consequently, we fall into**the state of commonly inflated, or deflated thinking** when we are unable to think straight and be consciously detached from computer programming**.**

***I repeat again that the process of developing true and aware digital intelligence must have aholistic outcome that works for the development of our higher consciousness.***

It should not be hampered by **the robot-like thinking** that becomes totally dependent on the machine because it shoots into the brain **DIGITALLY STAGNANT** habits. I witness mystudents' inability to ***get though the wall of blurry thinking*** that becomes a new educational challenge.

***If the flood of digital information is not properly digested***, it clogs the mind because it demands the multi-tasking approach to life that messes up the inner world the emotional and mental turmoil. Undigested intelligence results in the situations when a student does not assimilate knowledge; He
/she cannot ***disassociate himself / herself form the virtual reality,*** cannot feel mind-free to express an individual opinion, not the one that can be easily plagiarized and perceived as one's own.

***The digitally enhanced thinking skills*** need to be processed through the holistic **synthesis- analysis - synthesis** paradigm. *(See above).* To keep your consciousness intact, practice ***Active Auto-Suggestive Meditation*** and up-load your smartphone with a new inspirational tone.

## Use Technology for Your Self-Ecology!

# 8. Go Beyond the Basic Content of Your Intellect!

There is another side of the coin here. ***Despite extreme socialization, we feel lonely and self-un- reliant,*** seeking the solution of any problem by grabbing the cellphone and calling someone or getting in touch with a "friend" for Face Book. ***I think that technology makes us more self- reliant***. It's great to know that help is at a click of a finger, but let this feeling add security to your***independent being.***

A lonely, dependable person gets obsessed with electronic matchmaking; kids stop reading and become heartless and indifferent. We get irritable and emotionally insufficient due to **the lack of face-to-face and heart-to-heart communication** that we are hungering for.. Unfortunately, the unstoppable digital in-put and ***the cacophony of information*** that we get daily is often left unprocessed, unselected, unsorted out for its validity, with no aware attention applied to this indispensable process. As a result***, the neural code in the brain gets messed up,*** and it is very hard to untangle it, unless it is meditatively and consciously done.

So, we need to constantly ***develop our aware attention skills to oversee both the in-put and the output of our mental energy***. A rigid brain control should be taught to raise brain's adaptability tonew life demands. With the above said, the programmed living intelligence should not be appliedwhen the level of our curiosity is high, our rationality is self-controlled, and our wakefulness is conscious and ready to go forward on the ladder of evolution. **Be sane in every vein!**

Digital intelligence is amazingly creative when it is ***used consciously with much awareness*** and with a lot of sorted-out knowledge. These are the cases of. ***Steve Jobs, Bill Gates, MarkZuckerberg, Reid Hoffman***, and many more most brilliant advanced-thinking titans in electronicsthat have never rested on their laurels and that have bettered the world in so many business, cultural
, and social ways .

Only with the computers plugged into our **intellectually spiritualized volt** can we become ***more intellectualized, strategized, emotionalized, energized, better informed, and totally reformed!*** Digital equation with the neuro-linguistic self-programming that is being promoted now has to make us realize that every advanced idea sends a conceptual signal to the Universal Intelligence, and we are being fed back by it with the most amazing Aha moments.

We witness this interaction in the mind-boggling work of many computer scientists in any area ofknowledge that they explore. ***Let's stand in awe to their contribution to our evolutionary development*** and never forget about their incredible input into our lives. Obviously***, going beyondthe content of a person's own intellect***, and constantly expanding it is crucially important for a ***personality development and the preservation of an individuality***, inducting yourself with:

# I don't think in a Robot-Like Trance; I Relapse!

# 9. Digital Clicking in Love-Seeking

***Automatic thinking without awareness is the main dilemma of human existence now!*** Unprocessed digital intelligence deflates a mind-heart connection, if such clicking ever occurs! ***Numerous match services are booming, but heart + mind connection is swooning.*** Everyone is hoping for a soul mate, being unaware that **love-seeking should occur in five holistic levels**, too,starting with the spiritual probing of each other and finishing with the physical testing.

***The holistic perception of each other is the right thing that polishes the mind -heart ring!***

**Our love seeking should not be superficial and impulsive!** The framework of many sights fordating is set in scanning a potential partner by asking to fill in the questionnaires that are ***psychologically unprofessional and superficial***. They need to be holistically targeted in five dimensions of life, mentioned above. Digital engineering offers girl -humanoids for service, and it is, of course an amazing improvement of the sex toys. But such creations will impact the **heart**
**-to-heart** and **mind-to-mind** interaction between men and women, if not destroy love completelySex, like fast food, will become too addictive, and it will only generate more ***mind-heartdisconnection*** that we are supposed to repair. Too specified questions are not helpful here. The quality of our intelligence must be in an inseparable connection with the ***heart-based qualities*** ofkindness, compassion, care, and consideration for the other needs to be considered. Thinking + feeling awareness is the blueprint of self–creation and self-fairness.

## Only mind+ heart awareness builds up love fairness!

Being aware means getting rid of compulsive thinking and living in the past. Since our relationshipchances are limited to digital communication more and more, we need to use digital means to ***enhance, not deflate the mind + heart connection.*** . *Very soon robots will be able to duplicate and even surpass us in our unique ability to love. The key* to this strategy is our understanding of the power of the opposites and the ability to connect them in our behavior, *synergizing our joint intelligence and consciousness on the way.* **Self-change focused on emotionalizing the heart andintellectualizing the mind** is the solution in every area of our technologically enhanced life. -

In sum, to rationalize and humanize our lives, we need to learn ***the cause-effect strategy of livingand enhance the mind + heart connection.*** Transformation of consciousness to a higher level must be in the goal of our creative minds in technology. It presupposes the change of the frame ofthinking, on the one hand, and our observation of ***the cosmic Law of Cause–Effect***, on the other. Philosophy teaches us to constantly clean the dashboard of our life errors and start from a clean slate. Our kids should be taught to observe this Law at every level of their intelligence formation.

## "Eliminate the Cause to Change the Consequences."*(R. Wetherill)*

# 10. Raise Self-Discipline on the Digital Scene!

Digital intelligence is the greatest opportunity for us to establish the needed connection between the mind and the heart ***in the gaming business*** that is an open sky for us in this respect. At present,electronic gaming is the **field of war-mongering** and the display of super-human powers that develop aggressiveness, while the gaming opportunities are sky-bound in terms of helping us develop **AWARE ATTENTION** to professional intelligence*, or* the kid's / teenagers' / grown- ups' abilities to manage themselves.

Equally important is to use the artificial intelligence to cultivate good human qualities and lay thefoundation for a moral, ethical framework of the innovations in neuroscience and digital technology of the future.

We need to teach our kids and ourselves to love, care, display compassion, kindness, truthfulness,and love, through games, that often are heightening their aggressive, careless, and impersonal attitude to the outside world. ***Such games only generate inner imperfections*** and the disconnection with the parents because the participants identify themselves with the electronic personages and are losing their own individual characteristics. The virtual reality gets extended into the real life that is perceived as hostile by the brain that needs de-programming both mentally and emotionally. **We should better our inner world with the Artificial Intelligence aboard!**

Psychologists are desperately seeking the solution to the new addictive behaviors of the growingsouls that, when they reach their teens become unmanageable, disrespectful, and impersonal, seeking more fun in drugs, alcohol, and excessive, uncontrolled sex. (***Holistically applied awareness at*** every level generates a new quality of **PERSONAL INTEGRITY** that keeps a person from getting morally de-magnetized.

*5. Universal Awareness*

*4. Spiritual Awareness*

*3. Mental Awareness*

*2. Emotional Awareness*

*1. Physical Awareness*

We just need to be conscious of the changes that occur inside and outside us, getting much more aware of these changes and much more informed about the reasons that generate any emotional orpsychological consequences for us. **The Cause-Effect Law must always be in the Mind's Store!**

# Humanize and Harmonize Your Contaminated Inner Device!

# 11. Be Integral, Sincere and Kind! Be One of a Kind!

In sum, digitally enhanced **SELF-MENTORING** and **SELF-MONITORING** is a great opportunity for us on the path of raising our self-consciousness and a great help to our psychologists. Future avatars will be attuned to our psychological imbalances, giving us proper warnings, and helping us get together at the right time Start visualizing your ***spiritual, psychological, mental, emotional, and physical blocks.*** De-magnetize them and delete them witha conscious effort consistently. Freeze the unwanted patterns of behavior, emotions, words into a stone! **Be a Medusa Gorgon!**

Psychological awareness (*See Psychological Intelligence below*) needs to be introduced into the electronic games to widen the beneficial effect of the games that are based on the war-fields of aggressiveness and the demonstration of the super-power. There are boundless opportunities to engage players in the activities that develop **Holistic Living Intelligence Skills** and instill the noble qualities in them., like gratitude, and compassion.

**The abilities of being integral, sincere, and kind should reside in the Superhuman Mind!**

Also, we should use our future electronic friends, programmed into a wrist watch, or any other mind-boggling form ***to boost our spirit*** by suggesting the right decisions, expanding our educational horizons, and instilling in us the emotional control at the right place and the right time. ***"Be conscious. Consciousness mobilizes!"*** *(Neil Donald Walsch)*

**Shape your digitally focused mind to become more self-consciously refined!**

For sure, future gadgets will be properly tuned to our mental, emotional, and psychological states.Electronics is our step forward on the ladder of evolution, and it is the greatest device that we havehad so far in our exploration of the world.

***Marvel at the Internet that is God-set but be connected to the Conscious Life Net!***

To conclude, digital mind-programming will shape the brain You can program your smartphone not only to boost your spirit, but also to instill in you the qualities that you wish to enhance. Verysoon, ***we'll be able to fix our psychological and moral imperfections with the help of digital identities, called Avatars.*** Undoubtedly, **digital intelligence will promote our brain wavelengthto a much higher overtone** enabling us to heighten the level of our self-consciousness considerably with each evolutionary step of the humanity that will be even more mesmerizing withour future AI monitored extra-terrestrial travels.

# Wow! I wish I could live then in the Unanswerable When!

*(End of Stage One, Physical Dimension/ Digital Intelligence, division 3)*

# The Intellectual Odyssey of a Seeking Soul

*Vistas of Intelligence to install next:*

**Inflate Your Personal Pride with the Language Intelligence Might!**

| | |
|---|---|
| *10. **Universal Intelligence*** | **Self-Salvation** |
| *9. **Spiritual Intelligence*** | *( Universal Realm )* |
| *8. **Social Intelligence*** | **Self-Realization** |
| *7. **Cultural Intelligence*** | *( Spiritual realm)* |
| *6.**Financial Intelligence*** | **Self-Installation** |
| *5. **Professional /Creative/Intel.*** | *(Mental Realm)* |
| ***4.Psychological Intelligence*** | **Self-Monitoring** |
| *3.**Emotional Intelligence*** | *(Emotional Realm )* |
| ***2. Language Intelligence*** | **Self-Awareness** |
| ***1. General Intelligence*** | *(Physical Realm )* |

*/ Physical / Cognitive / Digital /*

***Language is the Portal of our thinking, speaking, feeling, acting, and perceiving.***

***Language is our Being!***

## To Make Your Life More Digitally Wise,

## Generalize -Analyze - Internalize - Strategize –Actualize!

# Stage Two

*Mini-Level of Intelligence* - ***Self-Awareness***

# Language Intelligence

***"There is only one language in the world – the language of God – the rest are just dialects"***

*(Umberto Eco "Universal Language")*

*(For more, see the book "Language Intelligence or Universal English"/ **www. Language-fitness.com)***

**Be Language-Fit to Feel Complete!**

**Have Language Awareness with Fairness!**

**Language is Me; Language is My Identity!**

# 1. Language is the Matrix of Our Thinking!

There is no intelligence of any kind without language in the brain! That is why ***Language Intelligence*** is at the basic, mini level of the holistic pyramid of intelligence**.** I have not started the presentation of **the hierarchy of intelligences** with the language intelligence because, unfortunately, before a person comes to the realization of the necessity ***to tame his tongue***, he firstgets interested in fixing his physical status with electronic gadgets at hand .

***Every language is a miniature universe and language intelligence determines our life in it.*** Eachlanguage phenomenon is rooted in the brain as the matrix of thinking and the root of our self- consciousness. As a matter of fact, "***pure reason***" *(Emanuel Kant*) is the Universal Intelligence by which every language phenomenon is roofed. Therefore, our language capacity development will always exceed that of the humanoids in the quality of our thinking and its conceptual display.

Language Intelligence separates us from the animal kingdom, and it gives us the main assert – our**THINKING CONSCIOUSNESS. "*Thinking is a subtle state of speech, verbal or written".*** *("Language Intelligence or Universal English," Book One)* **Rationalization of the language use is inseparable with the rationalization of thinking** that requires another level of **LANGUAGE AWARENESS** and **LANGUAGE CONSCIOUSNESS** that impact our Universal Being.

It is very important to remember that English, being the global language, is developing in people,using it, the **SECOND LANGUAGE PERSONLITY** of a very special, ***uniting character*** because it teaches people to use it for the global communication. Not only do we think in a new, digitally enhanced way, but **English also molds SL personalities**, adjusting the minds of the people who learn it to the new exponentially developing digital reality. Not to get into the life's gutter, instill in your mind the English mind-set: **"*Life is tough, but I am tougher!***

The words we use create the states we are in. Inspiring words inspire! ***Despairing words despair!***The most vital thing is to think for himself / herself. That is a big mistake of many parents that ***stereotype the thinking of their kids*** with their limited religious or national thinking. ***It is a mentally criminal act*** that is infested by school education, mass media, and the society overall. No doubt, **language intelligence permeates all the vistas of intelligence**. ***They are all in an inseparable celestial matrix of intelligence.*** The AI has already created different language models,LANDA and others, that demonstrate the conscious use of the language. It is a great stimulus for us to better our humanness with digitized language intelligence. Above, I have compared the interconnectedness of all the intelligences that we must develop on the path of self-formation to ***the Russian dolls.*** Like *the Top Doll* that encloses all the rest of the dolls in one structure, the **LIVING INTELLIGENCE** is governing our intellectual development and self-consciousness rising, acting as the **BLANKET INTELLIGENCE.**

# Language is the Core of our Cerebrally Conscious Existence.

# 2. Language Consciousness!

Unfortunately, **our language intelligence is suffering tremendously now.** The memory bank is clogged with dis-united information, the vocabulary stock of many people is appallingly limited, not just in English, but in their native languages that define personal thinking, to begin with.. Thereare many junk words and phrases that ***substitute the clear-cut expression of thought.*** The hierarchic nature of language mirrors the hierarchical nature of our consciousness, or unconsciousness. It is easy to place a person by the way he / she speaks.

## Language Intelligence is Language Consciousness in action!

The scope of our memory can retain ***seven or nine pieces of information*** or words at one presentation **(7+/ - 2 ).** The technological enhancement changes this capacity tremendously, but the brain won't be able to ever work at the speed of the hyper-computers that we develop. It is quite natural and understandable because the human brain has the capacity to choose.

## Say what you mean and mean what you say!

Since the *Artificial Intelligence* disciplines the use of the language, **we need to do the same**, but in our creative, conscious, and exponentially unstoppable way because we have ***flexible, not mechanically based language-intelligence***. Our interaction is based on self-consciousness, inseparable with *the Unified Field of Super-Consciousness*, that is perceiving our mental signals and responds to them through its digital means and our intuitive make-up.

***Language intelligence is" telling" the brain not only WHAT to think, but HOW to think, too.***

As a psycholinguist, I am sure that computers will never be able to beat us in our language intelligence and the flexibility of **thinking + feeling** capabilities. Computers can create art, compose beautiful music, inspire great minds of *Steve Jobs* type, but they will never produce another *Lord Byron or Tchaikovsky*, another *Einstein or Dostoevsky*, another *Sigmund Freud or Nikola Tesla.* The consciousness of a genius is inseparable with the unique language intelligence,***inseparable with the Universal Super-Consciousness***. Genius people have exceptional minds anda very rich emotional perception of the beauty of the world, that the artificial mind will never develop in the **HEART+ MIND** sync. Most importantly, *the Artificial Intelligence* will never master the cognitive and emotional intelligence holistically which is another step-in **LANGUAGE CONSCIOUSNESS** development, ***"inseparable with the vibrational character that our linguistic DNA is radiating."***(*"Wave Genetics" (Academician . P P. Garayev).*

## To Be Consciousness-Refined, we need to Intellectualize the Emotions and Emotionalize the Mind!

# 3. Language Awareness without Negligence!

Our ***Language Consciousness*** is part of the Universal Consciousness that will never be exceededby any machine because **no machine will ever be able to operate the language** the way our trained brains do. "***Language is a dynamic interplay of consciousness***." (*Academicians. A. Leontiev)* **Language Consciousness is promoting our evolution!**

Developing our intelligence, we are automatically raising self-consciousness, and both evolutionary accomplishments are impossible without the holistic development of languageintelligence, or the unity of **the language habits and speech skills** that determine our holistic thinking and can be considerably improved with more AI language development. The holistic thinking which is the work of both brain hemispheres can be instilled in our learners only based on ***the holistically developed language skills and the knowledge of the structural uniqueness ofthe language, especially*** used by the AI. The brain is a very complex machine, but its broad topography is symmetrical in managing of the two main parts of it.

The left hemisphere is *rational, analytical, and logical*; the right one is *instinctive, intuitive, andsystemic.* Language resides mainly in the left hemisphere that reasons sequentially, ***excels at analysis***, handles words, and ***operates the structural system of the language, or it's grammar,*** **the form** of the language**.** But language is in the right hemisphere, too. *The* right side of the brainreasons holistically, recognizes patterns, interprets emotions, and builds up thoughts and conceptswith the vocabulary tools that the left hemisphere provides. It is forming the conceptual meaning of our language interaction, **its content.**

**The form + content** language use defines our holistic thinking. That's why the development of the right hemisphere of the brain hemisphere is crucial. It is underused at present, and this fact undermines our holistic brain development. Obviously, ***knowledge of neuroscience*** in this respect, is essential for boosting our thinking abilities because driving a car blindly, we will eventually come to a dead end. **Language awareness boosts language intelligence** because is the pe- requisite for the cognitive development of the brain. A person who hardly ever reads and is tongue-tied in speaking cannot accomplish a lot in life.

***So, consciously developing our language intelligence, we are automatically raising self- consciousness that is the goal of our evolution.*** Mechanically programmed language cannot raiseself-consciousness unless the machine that is supposed to evolve the needed holistic **mind + heart**connection finds the way to establish it. The evolution is pushing us to this unity, and we can witness it in the best, most-advanced people among us. However, we cannot duplicate it *(speech* -**the right brain** + (*language* - **the left brain)** in a simulated mind with the precision of the Creator.

## The Whole Brain Intelligence is Our Evolutionary Goal.

# 4. Language Fitness without Cognitive Sickness!

We know that *the* ***digital neocortex is being created in us now***, but I think that, irrespective of thismind-boggling fact, we won't be able to raise our language consciousness that I have touched uponabove, without healing our ***Language Habits*** and ***perfecting the Speech Skills,*** both in a native language and English that is called by linguists "***Globish***" now. The polluted way the English language is used on the Internet and while people are texting on their smartphones is appalling.

**Language fitness is directly connected to RATION - EMOTIO sickness!**

The digitally enhanced evolution demands a new **LANGUAGE FITNESS** from us now! **LANGUAGE LITERACY** is violated, and language negligence is killing our holistic thinking development. New language-awareness needs to be instilled in our schools as early as possible because the damage that is done to our language intelligence might become irreversible.

**To evolutionary succeed, we must be language-fit!**

*(See www. Language-fitness.com / Excellence Award, 2019)*

The English language, when consciously used, can express absolutely anything in the most precise,thought-provoking, and insightfully beautiful way because it has **an exceptionally simple, lineargrammatical structure**. English is the first global language in computation because its structureis a one-way street. ***It's mathematically correct*** and, therefore can be used for programming. Theword-order of the English language lets the machine perform linear thinking operations. ***The lexical wealth of English has an incredible volume of a shot-cut scientific preciseness, too***. That's why there is never exact translation possible computer-wise from English into any other language and vice versa. The linear restructuring of the mind is needed when the English mentalitycomes into play. That is why English is rapidly developing from the status of a global language tothe status of the universal one, **the language of God!** Obviously, ***language operating robots should be programed by very good linguists, not just electronically smart programmers.*** We havea negative experience with poorly programmed dictionaries.

***"Our neocortex is virgin territory when the language mentality is created. It has the capacity of learning and therefore creating new connections between its patterns."*** *(Ray Kurzweil)* It means that if we learned the pattern consciously, we recognize it. The neocortex is continuously trying to make sense of the in-put, presented to it. So, we need to **constantly select the information,** and delete the one that clogs the neurological paths and blocks our creative thinkingand holistic brain development. So, consciously discipline your thoughts language-wise!

# Language is the Passport of Your Personality!

# 5. Language Pollution Impedes Personal Evolution!

**The lack of language controlling skills** is always connected with ***the lack of the bad language immunity*** that I have touched upon above**.** In the process of self-transformation, we cannot affordto speak casually, automatically, polluting our speech with banal phrases, junk words, profanity, and commonly used misjudgments. We must not let our bad language habits ***make us dis- empowered emotionally and mentally.*** We must kill them, or as Albert Einstein put it,

*"Bad habits have a good tendency; either you kill them, or they kill you."*

Good knowledge of a native language or of English is marked by our thinking, feeling, and actingin these languages. There is the flood of **LANGUAGE POLLUTION** on the Internet and aroundyou, but your language intelligence should not be affected by ***the avalanche of contaminated language***. There should not be any casual word use, without any monitoring of a thought. A beautiful British saying testifies to such a necessity.

*Self-Induction* **"*You never have a second chance to make a good first impression!*"**

The holistic approach to language-speech performance also demands that we develop a tight control over **the synchronicity between the language habits and speech skills** that we have noted above. It's not an easy thing to master, but it is not altogether impossible.

LANGUAGE HABITS **(*Pronunciation, Grammar, and Vocabulary*)** that constitute the language operandi system are formed very slowly, and they require aware attention to be paid to redesigning them **to** make our pronunciation or diction comprehensible, our grammar accurate, and the vocabulary use varied and exact in its transmitting our thoughts.

SPEECH SKILLS ***(Listening, Speaking, Reading, and Writing)*** are formed much quicker. Weneed to learn to be economical and exact in speaking, conceptually oriented in reading, and insightfully perceptive in writing. However, if the speech skills are formed on the automatic pilot,***without aware attention being paid to the process of conscious language use,*** bad habits are verydifficult to be eradicated. That's why many junk words that pollute the speaking of our kids are sodifficult to be rid of. **Bad language rituals result in a bad speech performance**. Bad speech performance results in a bad destiny because we stop paying conscious attention to the way we speak and to the exactness of the messages that we communicate with the help of the language. ***Bad language control and meaningless talkativeness generate*** family problems, relationship fractions, job problems, money loss, and, ultimately, health worsening. ***Only conscious, aware attention to the way you speak does the trick!*** *("Language Intelligence or Universal English")*

# Bad Language Habits Result in Bad Life Rituals!

## 6. Language Pollution Impedes Personal Evolution!

Since speaking is our first action since birth, it needs to be governed consciously all life. So, don'tever stop controlling your language, a native one or English, because the image that might have been once sharp and clear becomes dim and blurry with the lack of language control. What we donot use, we lose! **A minute of thinking is worth an hour of talking!**

Language degradation is a very quick process that drags us down unstoppably! ***Degradation in language intelligence is also connected with the degradation of a personality*** **-** slagging of character, weakening of will-power, emotional misbalance, and the **INFLATION OFTHINKING.** *(Cognitive Intelligence, Chunk 5)* So, motivate yourself to do any action by correctspeaking about it. Any problem gets resolved through **CORRECT LNGUAGE FRAMING** because any problem awareness is heightened by language exactness and the emotional control ofit. Consciously admitting the problem and your emotional reaction to it will help you deal with it.

**A soft answer turns away the disaster!**

The scientific effect of framing is well known in any area of life. **Language ecology + emotionalintelligence = the psychological** An unframed picture always looks less impressive. Now, professional language has a special framing, too, and it is, of course, justifiable because it serves as a short-cut to the brain in a professional problem solution. Unfortunately, we have depleted it with the **HEART + MIND** link, having substituted it with the **bureaucratically deaf language** that needs a special interpretation to be employed and that makes any relationship impersonal. Theregular, "***I don't care***, "I ***care less*** "come in this heartless place.

*Auto-Induction:* **I am much less if I am LANGUAGE- FRAMELESS!**

In short, the inner feeling of accomplishment that you have generated once should always be reinforced with your work on **LANGUAGE ECOLOGY** consciously., remembering the rule**,** ***Words may be verbal bullets. Don't kill anyone with them!***

*Auto-Induction:* **I Commit to Being Language-Fit!**

***Be devoid of an automatic speaking negligence.***

***Respect the language of the Universal Intelligence!***

**The Orchard will Bloom in a Blast if you tell yourself fast:**
**I CAN, I WANT to, and I MUST!"**

# 7. Language Correctness and Speech Exactness

***"Language is our first action in life!"*** *(A. A. Leontief)* There is no evolution of intelligence without ***language fitness*** because there is no development of consciousness without the clarity ofthought and the accuracy of word. Also, language has a profound effect on our life, and it needs tobe **grammatically correct and lexically precise**. We are defined by the way we speak and write!Language should always be in rapport with speech in the holistic unity of both hemispheres.

The moment a person opens his mouth and says something, you can place him for his social status,profession, attitude to life, and even his philosophy of life. Taking care of our physical fitness, orour physical intelligence, we often forget to control ***our verbal impeccability*** and, thus, we displaymost disrespectful attitude to the language in which we communicate our thoughts. In our digitaltimes, the adequateness of the thoughts that we impart and the language that we use have becomeappalling. We cannot establish ***a good mind-to-mind contact*** with a conversation partner in a face-to face communication or in the virtual space., unless the language and speech are in synch.

**We need to speak / write correctly, language –wise, and with precision, speech-wise.**

Correct speaking of a person relates to his correct behavior because it signifies to his correct thinking. Speaking with precision and, therefore, correctly, being able to adequately communicateyour ideas requires **a precise choice of words** that gives you the chance ***to say what you mean and to mean what you say.***

Obviously, **language unfitness always points to the lack of intelligence.** We should note here that any personal contact to be adequate must be based on proper language habits – ***good grammar,correct sentence structure, and a meaningful vocabulary choice.*** We must be sharpening our language skills all life to be able to express ourselves clearly and consciously, both verbally and in writing.

***The use of the language in the Artificial Intelligence programming must be especially accurate.***Language operating should be marked by scientifically verified, conscious, controlled, sense- making mode of speaking that requires also much richer vocabulary stock, too.

## The brain constructs the vocabulary of a single person.

It is always personal, and therefore, each of us speaks and thinks differently, and this difference is mentored by the richness / scarcity of the vocabulary that we use. It will not be an overstatementto say that there is no personal growth without a substantial growth of the vocabulary stock in a person's brain. (See *"Language Intelligence or Universal English!"/2019*

# The Richer the Vocabulary, the Better the Thinking!

# 8. Practice Visual Listening with the AI Mind-Reforming!

We can diversify our language habits and speech skills much better if ***we read consciously***, payingattention to what we read, **conceptualizing the text while reading or compartmentalizing the knowledge**. It is also vital to pay aware attention to the richness of the vocabulary that is used bythe writer. These reading skills should be developed in schools, teaching the extensive ***mind- provoking reading -*** analyzing, interpreting, and synthesizing the information verbally and in a written form. The skills of speedy reading that I have referred to before are essential here.

### Synthesis- Analysis –Synthesis!

**Professional vocabulary skills** are also extremely important in this respect *(See Professional Intelligence)* The professional use of the language means that we need to make the right vocabularychoice to express our thoughts adequately in any professional environment and to describe any problem in a concise way and to find a short-cut to its solution as quick as possible.

*Self- Induction:* **"Talk the Talk!"**

This rule is applied to any area of expertise because every profession has its own vocabulary stockin use. If we are not aware of this vocabulary, our talking about job problems is vague and, hence,the decision-making skills will never be formed adequately.

***It's also very helpful professionally*** *to visualize the professional terminology.*

In sum, evolution of a language use requires ***the paradigm shift in language competence*** becauseit redirects our attention from the predominant focusing on the speech skills to a holistic and keenlyconscious linking of ***language habits and speech skills.*** *(Chunk 2 above)*

## To Life-succeed, be language habits and skills fit!

We can tell our brain not only what to think, but how to think and what to say! Going along this path of perfecting our language intelligence, you will inevitably stop being self-conscious especially when speaking in public and ***become more language conscious***. With your **AWARE ATTENTION** paid to the language in you, you will be much better perceived by people in your **face-to-face, mind-to-mind, and heart-to-heart** communication. *(See Language Intelligence orUniversal English, Book 3)* The communication at the heart-to-heart and mind-to-mind basis is an opportunity to be changed by another person!

# Stop Seeing Yourself in the Language and

# Start Seeing the Language in You!

## 9. Language-Speech Immunity

To conclude this part, language intelligence mostly depends on your **LANGUAGE-SPEECH IMMUNITY** against using bad words, junk words, cursing, insulting people, yelling, constant lying and, most importantly, judging yourself and others. ***The ability to control thinking and speaking right is a real might!*** Bad language-speech immunity is the obstacle to **SELF- REGENERATION** overall. Language pollution is polluting the mind, and therefore, a person's self-consciousness.

*Auto-induction:* **Watch your thought to control the word!**

Regrettably, bad language habits get harbored in the brain, and they require **individual, intentional, and conscious surgery** to be done to irradiate them from there. In his wonderful book "***Martin Eden***" by a great American write ***Jack London***, the main character manages to become a famous writer thanks to his tireless work at his language skills that were polluted by hisformer life of a simple, uneducated sailor. *Jack London* writes, ***"Nobody can change human nature, but we can change people's thinking and speaking and though that their behavior can be changed, too."***

**Language de-contamination and its eventual healing is a willful process!**

Try to have a small language survey of people's speaking around you. Go for a walk in the park or just sit on a bench and watch the world go by, focusing especially on how people are talking behind your back or in front of you when passing by. You will be appalled to discover that they are mostly talking about someone doing something wrong. Negative assessments of people and life are heard all over: ***"That's absurd. / It's insane. / It's crazy. / It's ridiculous. / I can't take itanymore! / Are you nuts? / Leave me alone. / You know what I mean, etc.*** There is a lot of cursingaccompanying these reactions, too.

I am extremely grateful to ***Oprah Winfrey***, **the woman of a mission,** that had shaped the minds of millions in the world and injected eternal, positive values of love, respect, and morals into her public. It was the first TV show that I was very much impressed with, being very used to the Russian style of intellectualizing the public through TV.

People hunger for the shows of this level of intellectual challenge and emotional pumping. **Whata great chance to obtain professional and personal magnetism and inner integrity!** No doubt,the quality of thought generates the quality of speech and modifies behavior. We need to share our most inspiring emotions, talk about great things that we have done, complement each other and oneself for the actions that went unnoticed and unappreciated.

# Intelligence and Attitude are shining through the Language!

## 10. Not to Cast Shadow on the Sun, Tame Your Tongue!

In sum**, *"Language is impregnated with individual characteristics of a person"*** *(John Binge)* People who are language–savvy get along better with relatives, friends, and colleagues at work. Their thinking is normally well-structured, and they can better communicate their thoughts to theirlisteners. They think before saying something, and they oversee their tongues. ***A language-savvy person has a strong and confident*** **LANGAUGE PERSONALIYTY**.

**Goal # 1 – Develop a new language personality- the primary goal in working on language intelligence.** A language intelligent person manages his language as much as he manages himself.Every religious book of wisdom tells us about the absolute necessity ***to tame the tongue*** since uncontrolled speaking is sinful. In our knowledge–based and increasingly competition–driven society, ***language skills are certainly number one aptitude*** and determines our intellectual health.

*Auto-induction*: ***"Be impeccable with your word!*** *(Don Miguel Ruiz)*

**Goal # 2 - Language Intelligence means thinking first and saying something next.** If you become polished in your language in tone and vocabulary, if you are very outspoken and clear-cutin self-expression, if you can ***express yourself clearly the first time around***, your language intelligence is healthy and solid. All these qualities become attainable if you manage your language, or, in other words, *if* ***you put the letter" T" before the letter "S"*** in your language use.

*Auto-Induction*: ***Think before speaking!***

**Goal # 3 - Redundancy pollutes our communication and kills our precious time. *Speak less*!**Language mirrors your personality, and it helps establish and maintain a better rapport with peopleif you are clear, concise, and exact in self-expression.

*Self-Induction;* ***Say less! Do not pollute your thinking space!***

**Goal # 4 - Learn to hold your tongue to avoid saying more than you can!** Language communicationneeds an absolute establishment of a ***mind-to-mind*** contact with an interlocutor, especially now, at the time of digital communication. Only conscious and concise communication teaches a person***to recognize the relationship within the structure of thought.*** Only then can a person end up saying what he / she means to be understood by your conversation partner adequately.

*Auto-Induction:* ***Do not verbalize everything you want to say!***

**Goal # 5 – Say less but mean more! Learn to use and decode implications!** *To* hit this goal, you must enrich your cognitive culture by ***rationalizing your thinking and speaking.***

## Be Language-Fit to Succeed!

# 11. Avoid Speaking with Nothing to Say!

To conclude, we desperately need **RENEWED LANGUAGE SKILLS** for our new electronicallybased and choice-driven economy. Speaking and writing with implications is a great gift! The Artificial Super Intelligence requires **the economy of the language** and

**SCOPE OF THOURT! That's our new, digitally fortified FORT!**

***"The method of the tip of an iceberg"*** was introduced in the world literature by a great Americanwriter *Ernest Hemmingway*. The tip of the message, like the tip of an iceberg that is seen over thewater, is said, while the rest of the iceberg or the body of the message that you want tocommunicate is under the water. ***It is implied, and it needs decoding.*** Such communication is very economical, but it necessitates a great deal of awareness of the subject of the conversation. ***This is what is required in the professional intelligence development nowadays.*** (*See Professional Intelligence, Stage Five)*

**Goal # 6** – **Language respect is also self-respect!** It has become a bad habit with many peopleto assess things negatively and to complain about being discontent with the state of things in theirlife, or the life of other people. ***Judgmental speaking*** has become a norm among young people, too. Every other word in their conversation is ", ***She is like..., he is like....***" behind that person'sback. Listen to the people talking around. They try to impress their listeners with something out of line, judging and condemning someone or something, expecting a listener to back their judgment up with the same train of thought. Regrettably, the response is, "***I Know what you mean...***" *or* "***Tell me about it.***" **Soul-compromising, we are not wising!**

**Goal # 7** - **"Clever people talk about ideas, mediocre people talk about events, stupid peopletalk about people."** *(Rubaiyat of Omar Khayyam)* Judgmental speaking is being instilled in us bymass media that focusses only on the negative side of life for its sensational value. Why don't welearn to notice the positive, constructive things first, Also, **the uncontrolled language behavior** of people in some TV shows and their manners are accepted by the public as a norm. The languageof the participants of such shows is often derogatory, and ***the inflation of their*** *language* ***skills*** signifies to the inflation of their thinking that's hardly curable. Nevertheless, such tasteless and harmful shows are appallingly popular. ***Imitations, not the number of viewers, is degradation!***

**You and the Language in You are the Inseparable Two!**

*For more on language fitness, see the books "Language Intelligence or Universal English"*

# At the Universal Bay, you are What You Think and What You Say!

*(End of Stage Two- Language Intelligence)*

# The Intellectual Odyssey of a Seeking Soul

***Vistas of Intelligence to install next:***

**Be a New, Intellectually Spiritualized and Emotionally Diplomatic You!**

*10. **Universal Intelligence***
*9. **Spiritual Intelligence***

**Self-Salvation**
*( Universal Realm )*

*8. **Social Intelligence***
*7. **Cultural Intelligence***

**Self-Realization**
*( Spiritual realm)*

*6. **Financial Intelligence***
*5. **Professional /Creative/Intel.***

**Self-Installation**
*( Mental Realm )*

***4. Psychological Intelligence***
*3. **Emotional Intelligence***

**Self-Monitoring**
*( Emotional Realm )*

**2. Language Intelligence**
***1.General Intelligence***

**Self-Awareness**
*( Physical Realm )*

*/ Physical / Cognitive / Digital /*

***Study the Emotional Diplomacy Skills and install the AI refills!***

**Our Emotional Fitness is at play today!**

## To Make Your Life Intellectually Wise,

## Generalize -Analyze - Internalize - Strategize – Actualize!

# Stage Three

*Meta-Level of Intelligence* - ***Emotional Dimension*** – ***Self-Monitoring***

# Emotional Intelligence

*Inflate Your Personal Pride with the Emotionally Controlled Might!*

*Change the Internal Ritual-*

**"Drink the Juice of Life, but Do Not Drown in It!"**

*(Sadhguru)*

# Emotional Intelligence with Much Negligence!

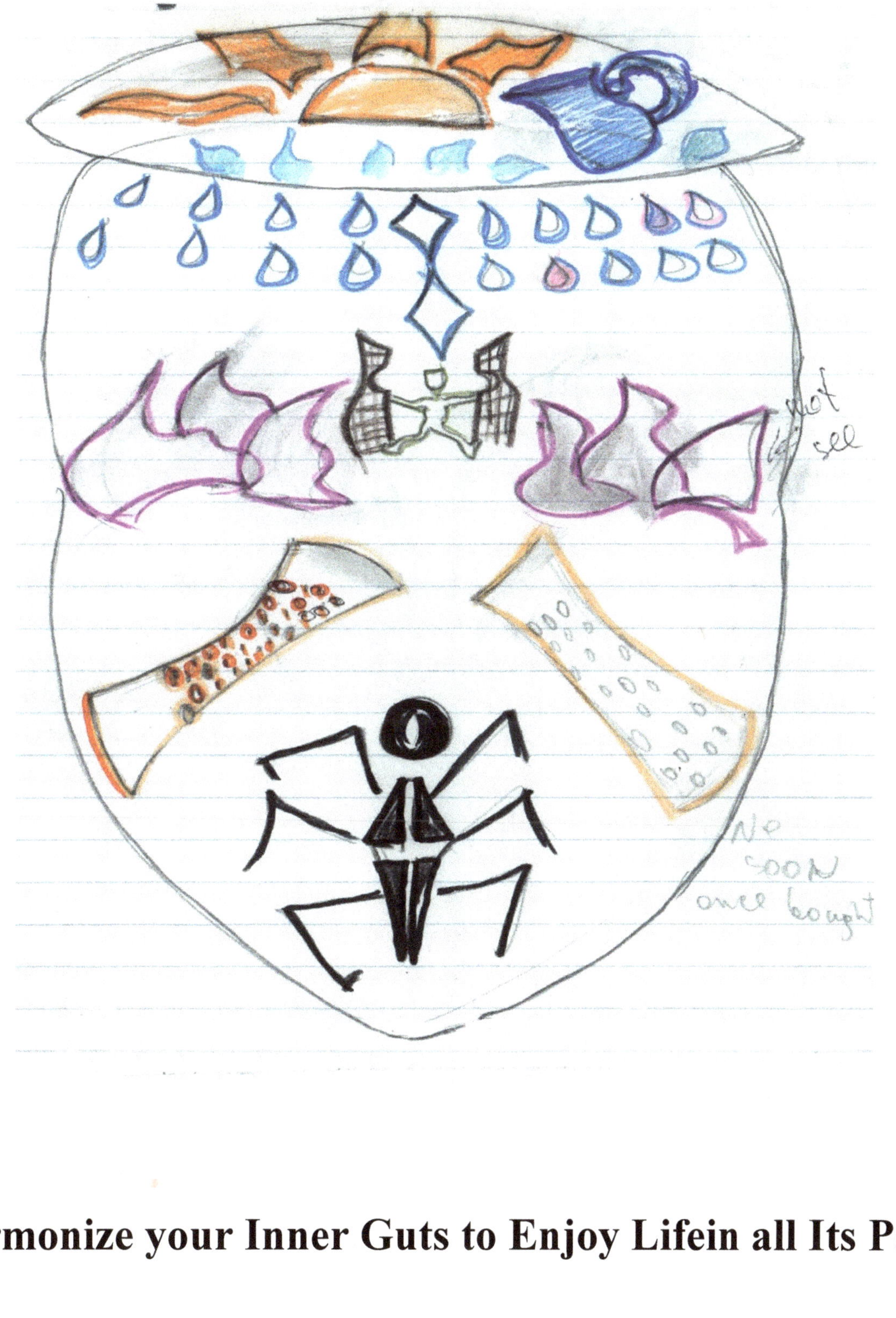


## Harmonize your Inner Guts to Enjoy Lifein all Its Parts!

# 1. Life is Mentally Emotional!

Science proves that ***life is mentally emotional,*** and the present-day state of our emotionalintelligence shows the worldwide trend for the people to be more troubled emotionally, more depressed, more impulsive, and aggressive. "***We are ruled by the one who angers us."*** *(Lao Tzu.)* ***One absolutely needs to out-power his emotions with the mind!*** This restriction uplifts consciousness because we heighten the level of internal commitment and obtain emotional wellness that is inseparable with physical and psychological well-ness that is in your ability to put the heart and the mind in sync and learn **the ABC of EMOTIONAL DIPLOMACY!** Your spiritthat is in the center of ***the spiritual fractal*** connects your heart and the mind and the integrity of the whole fractal depends on it,

**Body + Spirit + Mind + Self-Consciousness + Universal Consciousness!**

Studies reveal that success takes more than intellectual excellence. We need to develop the skillsthat raise us above the crowd. These are the **HOLISTIC EMOTIONAL SKILLS,** acquiring "***emotional intelligence'*** *(Daniel Goldman)* that is the next most important intelligence after language intelligence and the one artificial intelligence has to master. *Auto- induction:*

*Self-Induction*: **Intellectualize your emotions and emotionalize the mind! Be One of a Kind!**

It means that you put your ***mind over the emotions*** *(Introduction, Thesis 6)* and think before actingon the one hand, but you are also aware of the reason of our emotional misbalance, on the other. Unfortunately, in the hierarchy of our consciousness development, ***the cone of emotions is over the cone of reason.*** We are not in charge of our emotions yet. We still tend to be more animal- like in our reactive state of mind, rather than human-like in a responsive state of **conscious mindcontrol**. Dysfunctional thinking is the result of negative, uncontrolled emotions.

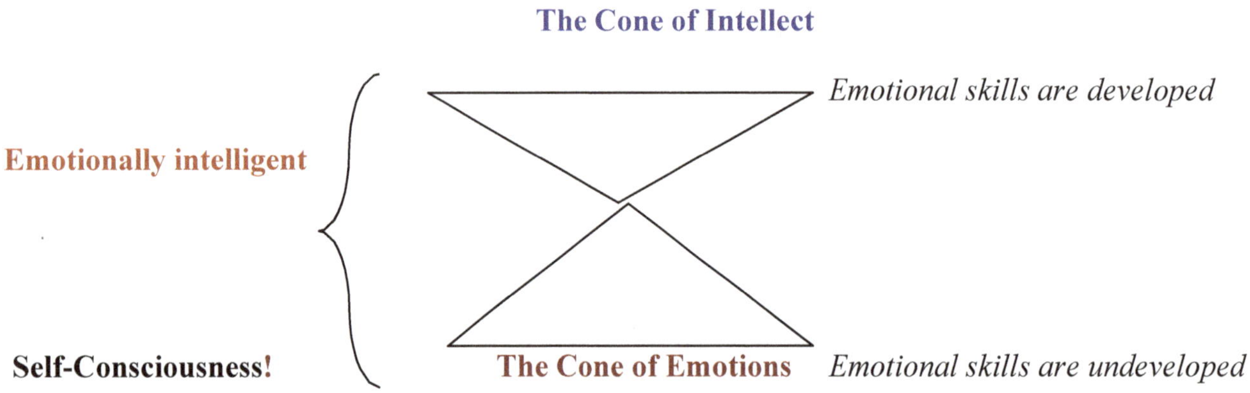


## You are what you choose to be in an emotional turmoil – a Lady, a Diplomat, or a Rat!

# 2. Emotional Intelligence without Negligence!

To obtain ***Emotional Intelligence*** of this sort, the most important skill to develop is **to consciouslybeat your impulsivity and learn to remain calm in any situation.** It is an accomplishable taskif we learn to install the **HEART + MIND** link into our being. The term *Emotional Intelligence*points to the essence of this unity that we need to consciously establish .in every life situation.

Our problems start from early childhood. When a child is small, the parents treat him /her with **theheart and mind in sync**. The disconnection in the heart and minds of the kids and the parents starts with the puberty of a growing personality when the common sense of the parents is subordinating every move of his / her self-expression. ***The aura of the vibrational energy changesin the family,*** and it resonates into the loss of this essential unity in both.

As a matter of fact, our relationships start and finish due to our following either the heart or the mind, not both, in their unbreakable unity. The fundamental ***Law of Cause-Effect*** cannot be reasonably followed in **the disconnection of the heart and the mind**, and the consequences thatwe face in every area of our common life are usually appallingly.

Every day, an emotional stress becomes a new reality that requires new tools of coping with it. Sometimes, stressful events come so fast that there is no time to recuperate emotionally. We become irritable, impatient, messy, and irrational in our thinking and often stop bothering about self-development or self-realization altogether So, becoming emotionally-intelligent also means that we become aware of the fact that we have an ***emotional and rational brains*** – the limbic system and the neo-cortex, and that these brain systems interact with each other. Obviously, beingemotionally disciplined requires a lot of self-work and self-control, self-knowledge, and self- awareness! "***Why are you always anxious?*** –*asked Jesus his disciples*. ***Can an anxious thoughtadd a single day to your life?***"

True, in certain situations in life, when the emotional brain becomes predominant, managing emotions is especially hard. When we are emotionally distraught, a quicker signal is sent to the ***Amygdale***, which is the emotional center in the brain. At such moment, we fight, yell, and argue because we completely lose the ability to be rational. But if we focus ***our aware attention on themind,*** it gives the command to the brain, and we get emotions in control.

*Mind* ⟹ *Brain* ⟹ *the Rational You!*

*Unconscious* ⟹ *Conscious* ⟹ *Superconscious!*

## Only with the Heart and Mind in Sync can we Stop Being at a Constant Conflict Brink!

# 3. Emotional Self-Awareness

***Emotional self-awareness is essential to gain for your emotional self-quest***. You must be awareof your **emotional keyboard**, and what keys you, not anyone, must push so you could sound ***Pianissimo or Crescendo***. The problem is that you must be a self-playing instrument, not an automatic piano that is being coded for a certain melody by the brain.

**Emotional attuning is like tuning a musical instrument's refining!**

A good musician would never let anyone do the tuning up of his instrument for him because no one is able to hear his tool from within, and no one can operate it in lieu of the instrument's specific inner sounding. **Emotional awareness means you are the pianist, not the piano!** Consciously assessing your emotional state, you can control it and monitor it the way you need. ***Confucius*** considered the feeling of shame to be the strongest indicator of self-awareness and the best motivefor self-improvement. ***"If you do not feel shame for what you did, you will never improve."***

**So, conquer your mouth with your inner wows!**

We also need ***to develop the ability to use emotions to enhance thought.*** Using ***the Auto- Suggestive kinesthetics,*** you can consciously stop the unpleasant emotion and maintain a positiveattitude even when we experience difficulties. Don't rely on anyone to give you a hand and motivate you. Emotional awareness generates **EMOTIONAL DIPLOMACY**!

*Auto- induction*: **To Be Inspired, Be Self-Inspiring!**

***Emotional awareness*** is the main boost in life because *we* ***first behave joyfully, then we feel joyful!*** Smiling and laughing, we are giving our brain an oxygen shower, on the one hand, and weare shaking the cells from within, letting them get rid of the pollutants, or mental- emotional residue, on the other. Interestingly, science proves that our facial expression take place before weexpress emotions, not after the emotional outburst. **Learning to be joyful is learning to be conscious** We read a lot about the necessity ***to smile to feel happy, not smile because we are happy***. However, the ability of seeing the seeds of contentment and joy is yet to be developed by every one of us consciously. ***We need to put the mind and emotions in sync. Feel but think!***

*Edgar Cayce* suggests using the command to the mind," **Halt!**" to help oneself stop the negative expansion of emotion and put the mind in command. This inner command gives you a chance toswitch gears. Put a note on the windshield of your car:

**You cannot say, "More Sun!" You cannot order, "LessRain!" You Must Sustain!**

# 4. Emotional Skills and Techno-Babies

The future of humans does not look promising as it encompasses the accelerated development of Artificial Intelligence with ***artificial love that is without the spiritual content.*** Every student of mine says he / she believes in God, but what it means in the actual sense, none of them can clearlyexplain. The difference between being religious and being spiritual is very vague in their minds., but gaining **the spiritualized intelligence is our universal goal**. *(See Mission Statement above)*

A new generation is raised on abstract, unspecified information that they get quickly adjusted to and acculturated by. ***Children are raised as techno-babies*** that get attracted and adapted to the new technology very quickly. Of course, this pace of development being very important evolution-wise, has a bad impact on the kids' brains and their nervous systems because we do not consciouslychannel this process in the way of developing our own and our kids' ***Emotional Intelligence Skills***.Children get emotionally overloaded and therefore, are often rebellious. There is no flat ground, no security, no rest, and no compassion if they slip down. Obviously, developing our emotional skills should start as early as possible, and ***it should be studied as a separate subject at school!*** *Daniel Goldman's* book "*Emotional Intelligence*" ***was*** published more than ten years ago, but its wonderful ideas are not put to practice in our schools yet.

**The Emotional Diplomacy Skills must be put on the character-driving wheels!**

Today, young generation is running on ***a trade mill that is set on the uphill mode,*** but, unfortunately, it takes them down intellectually and spiritually. We are all in the same boat. **The world has become very impersonal and lacking in emotional equilibrium**. There are sex or age differences in emotional intelligence, and our competence and incompetence in them keeps growing throughout life The sense of measure is hardly instilled in us or in our kids.

There is one more thing to consider here. A badly needed **mind-to-mind** and **heart-to-heart** mating. It is possible only" ***if we return to the only place of power- the present moment."*** *(EckhartTolle)* Dealing with the present emotional trauma then and there does not leave the residue of the past emotional memory. A beautiful quotation from a great novel "***Love Story***" *by Erick Segal* isa great example of such a situation: ***"Love means never having to say Sorry"***

Regrettably, we watch movies and TV shows in which cakes are thrown into faces, wine glasses are emptied onto an offender's lap, emotional outbursts and ugliest reactions are followed with storms of phony laughter. The tasteless, **lacking in the sense of measure** shows, primitive jokes,and endless commercials may be fun once, but watching them all the time makes us dumb and characterless. ***What's worse, they develop the habit of killing the time senselessly***.

# The Sense of Measure is Our Treasure!

# 5. Emotional Impeccability

The skills that we need most in developing emotional intelligence are the skills for the emotional ***equilibrium, stability, and conscious self-control.*** These noble traits of character are instilled in kids of nobility in every nation and country of the world. But such training must be introduced inregular schools that will be raising **EMOTIONAL ARISTOCRACY.**

Interestingly, different interviews with humanoids demonstrate their self-awareness and a great emotional control. They are aggressive in their intentions to destroy the humanity, but they do notdisplay anger, irritation, discontent, etc. Such neuro-algorithms were not programmed in them. Butthey are deeply engraved in us, and these "***bad habits need to be killed, or they will kill us.*** *(AlbertEinstein).* There are also many TV programs that promote yelling, fighting, pushing, verbal disrespect, aggression, and other animal-like behavior patterns. What can we expect from our kidsand young people? They should at least know the meaningful succession of the head, the mouth and the fists on the hands, telling them," **First think, then say, then do**!" *(The Law of the Right Human Behavior)*

Why don't we follow or show the wonderful examples of ***the Schools of Ethics*** in history – the school of gentlemen and ladies in Great Britain, the school of nobility in pre-revolutionary Russia,or ***the Spartan system of upbringing*** that stressed love of intellect and knowledge as much as physical exercise. Sparta was also characterized by an unusual degree of self- government, freedom, and responsibility. These schools instilled **gallantry, reserve, manners, respect, and other great values in their learners.** We still feel their contribution to the world in raretestimonials of such behavior on the part of the **IMPECCABLY MANNERED** men and women.So, the etiquette, manners, morals, polite rituals, and restricted talking must be studied and **constantly self-guarded** in every aspect of your life, especially your love life that affects our emotional being more than anything. Mind it, please, the more love partners you have, the more turmoil you generate in your life. **To have love, be love!**

In sum, ***good manners and a reserved attitude to problems*** have been instilled in people for centuries. Why not have them back and consider the necessity to oversee emotions a priority.

*Self-Induction:* ***A posture, a smile, and a good mood are my emotional food!***

With the ***Artificial Super Intelligence behind this aspect of Emotional Intelligence***, we will no doubt succeed more in enriching our kids' education and expanding their outlook immensely in this respect. Researchers recommend that we first consider the capacity to perceive emotions in yourself and others. So, let the well-known saying below never lose its freshness***." If you don't have something nice to say, don't say anything at all!"***

# Digitally Power Up Your Inner Rehab!

# 6. Self-Assessment and Self-Monitoring

To empower yourself emotionally, get down to the business of self-assessing your own ***physical, emotional, mental, spiritual, and universal potential.*** Emotions and thoughts become de- personalized though awareness. The knowledge about oneself is the starting point in self- formation. Therefore, holistic self-**assessment is a priceless tool for emotional awareness.** Auto-suggestive work that you should resort to at every level of your intelligence development will bemore effective if ***you keep assessing your self-growth at every stage.***

Defining for yourself what you want to accomplish at each level separately - **Mini, Meta, Mezzo,Macro, and Super**. The universal level of self-control reminds you of your God -given goal that helps you prove your ***personal exceptionality*** at all the other levels of self-growth.

**"The biggest mistake in life is not to correct your mistakes."** *(Confucius)*

Such self-assessments must be done without ***criticizing and victimizing yourself*** and with a lot ofself-love, self- respect, and self-consideration. *(See Self-Love Meditation, Physical Intelligence)* All emotional changes happen from within. The latest research shows that our DNA is affected by inner and outer vibrations, the verbal presentation of our thoughts and behaviors. A person gets **demagnetized and de-humanized** due to these changes! A great Scottish poet *Charles Mackay* in his unforgettable verse*" You Have no Enemies, You Say,"* wrote,

***"You've never turned the wrong to right,***

***You've been a coward in the fight!"***

Having assessed yourself most objectively in general terms with the objectives outlined above, you may get down to a specific **EMOTIONAL REFINEMENT in** *a particular dimension- physical, emotional, mental, spiritual, universal.* Test your attitude in each of the levels, starting with the universal one-***your dedication to your goal*** that is supported by your God-given exceptionality, discovered and protected by you. Thus, your **SELF-REFINEMENT** must alwaysbe based on your **SELF-WORTH.** Interestingly, self-assessment starts playing a significant role not only in psychology, but in business, too. People with the ability to objectively assess their owntalents and abilities are in great demand now because ***they sell their skills more adequately and***, therefore they are more marketable, competitive, and compatible. **The humanoids must be trained to be self-reflective and do self-assessment, too.**

In sum, you must work on the objective vision of yourself for your own good, not for anyone's appreciation. True, a supporting word backs up our self-esteem, but it is not propelling the growthof self-worth. *(See the book "Self-Worth", New York, 2019)*

# It is Hard to live with Self only if you do not Know Yourself!

# 7. Cleanse Your Emotional Memory!

We all experience the moments when we say, "***I can't take it anymore!"*** and "***I forgive, but I don't forget.***" Studies of emotional memory lie at the heart of the brain science now. It is the emotional memory that is prompting our impulsive actions, making the habits of uncontrolled emotional reactions very hard to break, and the memory that is responsible for making the heart- breaking insults and humiliations unforgettable. ***"What makes the brain a brain is memory, and emotion is arguably the primary sculptor of memory."*** *(Scientific American, May/ June 2016)*

Yes, **it's easier to forgive than to forget!** It's impossible to forget unless ***you do it intentionally***,erasing emotional scars in your brain ***consciously*** and never putting salt on the old wounds again.Just send a bad emotional memory into the recycle bin of your brain with a willful move of your mind. To be able to do that, **you must forgive yourself first** for having let somebody hurt you dueto the lack of self-worth and the absence of a true protection of your soul**.** Our inner self, our **INTUITION** and **CONSCIENCE** are our direct lines to the Universal Intelligence, and they always on guard. Regretfully, we do not pay aware attention to their signals. We are negligent of those signs in a rush to react at an impulse.

*Self-Induction:* **Listen to yourself! Take your own council first not to react in a reverse!**

I have already reminded you of the old schools of gallantry. From ancient times, such schools prepared young people to oversee their emotions, putting the mind in command of their lives andobserving **EMOTIONAL DIMPLOMACY** in a person -to-person communication**.**

*The movie "**Secret**" and the book by **Rhonda Byrne** are a great contribution to our self-growth inthis respect!* We also love watching old movies that depict the people of nobility and honor, real gallantry and manners, self-worth and mastered self-reserve. A great American movie" **Meet JoeBlack**", with two amazing actors, ***Anthony Hopkins, and Brad Pitt***, communicates to us many thought-provoking messages how to build up inner integrity and have power over death. ***It is a real manual of Self Worth and a great piece of art!***

Apparently, consistent auto-suggestive inductions serve as ***the best mind-fits and soul-fits*** from literature, art, music, movies, and wonderful self-help books because they break bad patterns andinstall good ones ***through the process of intentional and conscious synergizing the mind and the heart.*** The humanized machines now can demonstrate an array of different emotions at a command of an operator, but they should also be trained to help us control our emotional outburst by tuning in to our neuro-link perceptibly and helping us establish self-control.

## Follow the Intuitive Guidance of Your Mind and Sensations with timely Digitized Patience!

# 8. Sincerity is the Best Emotional Dexterity!

***The inner light of a person is always tested with conscience and sincerity***. A wide sincere smile,for example, will detonate the strenuous moment and help you ground the negative impulses in yourself or someone that had generated negative emotions in you. ***Christ's instruction not to strike an offender back is a great exercise for the mind and willpower*** that we must practice consciously. Your self-worth won't let you belittle yourself to the level of the offender.

*Auto-Induction*: **Be kind to the unkind. Be One of a Kind!**

It is still a very controversial statement for many people who want to justify their ***"eye for an eye"***philosophy, considering the emotional reaction to be natural in this respect. However, the responsethat ***Christ professed needs a lot of conscious action to be put to work,*** when you are in control of your mind-power. Such **SELF-EMPOWERMENT** will help you stop victimizing yourself andfill yourself up with self-pride and honor. Thus, you'll manage *to defy a low-gravity common sense impulse* that you might get. ***What a feeling that is!***

*Auto-Induction:* **Those that defy the common gravity - fly. Those that crawl - die!**

A powerful image of yourself in the mind, ***the URV*** *(the Ultimate Result Vision)* of the best of you*( see Physical intelligence, Chunk 5)* is very helpful in this respect, and ***it should be visualized continuously and consciously.*** Teach yourself self-empowerment while *thinking, breathing, eating, talking, exercising, working, driving, walking***,** etc. **Learn conscious living!** You are performing self-cleansing or **SELF-RECYCLING** without which the clogging of your emotionalmind will happen again and again. Therefore, conscious work at **SELF-ACCULTURATION**- might be either stopped altogether, or your uncontrolled weakness will fling you to the zero stage.

**Beware, emotions need as much order as thoughts do, even more!**

Apparently, you need to come to terms with your inner self because life and relationship are a variable reality, not an absolute one. It is only natural that women start cleaning, cooking, or putting everything to calm down when they are very upset. This practice helps them discipline their emotions and the sense of **SELF-WORTH.** That is why a woman's brain is more developedintuition-wise. In a wonderful book, *"****The Flower of Life,"*** *Drunvalo Melchizedek* discloses the ancient secret of the flower of life (**Merkabah**) that represents ***the Sacred Geometry*** of the unity of the heart and *the* mind. It is the magnetic center around the heart, its essential core that ***encompasses the entire human system holistically*** and enlightens your entire being. Humanoid will never have it. It will remain sacredly human, and our goal is to charge it with more true love and light, questioning ourselves, " ***What is the nature of my life? Why is it incomplete?***

**"Be the Light onto Yourself! Be Yourself!"** ***(Buddha)***

# 9. Emotional Diplomacy, Tolerance, and Patience!

Being emotionally strong for oneself and the people around us is the demand of life that is becoming emotionally tougher and tougher with each new century. ***"Mastering emotional intelligence is a demand of our time."*** *(Daniel Goldman).* Each of us needs to have a diplomatic key to operate some one's psyche. If we have it, we develop a solid friendship with that person foryears on end. So, keep your friendships on Earth and never lose their supporting force.

All we should try and **adjust the key to a person's heart and mind with patience** and a great deal **of EMOTIONAL DIPLOMACY.** As a result, we will develop ***emotional competency*** that will back us up in any emotionally charged situation. *(See my book "Emotional Diplomacy")* ***Emotional competency is how well we handle ourselves and others.*** You are not a passive receiver of feelings. **You are the creator of them!** You must take responsibility for the creationof your emotional will that is the electric line to your spirit. Do it patiently and consciously.

*Auto Suggestion*: ***"Will your life more!****"(Carl Yung)* ***That's the Law!***

Once you start to interact with another person, you put your emotional diplomacy to a test. In fact,it begins before that moment. It begins when you first open your eyes in the morning, when you first interact with yourself and when you say "***Hello***" to a new day. Saying a happy hello to yourself every morning must become a habit! ***But beware of anyone's bad mood around you. Don't get injected by it!*** *Use your* **SELF-GRAVITY** to ground any negativity then and there.

I admire Americans for their ***happy Hellos!*** It is so pleasant to see a smiling face everywhere, to be welcoming anyone while holding a door for him / her. The Art of Living lies in such light - filled moments, in our conscious choices to be happy, joyful, grateful, helpful, and smiley. Choosethe state of mind of ***a life–lover, a life-winner, and a life -appreciator. Thank the Sun, the God,and the world around you!*** Trust your intuition at the time of an emotional friction! Feel the **NOW** of your existence because you can only imagine tomorrow or feel it coming. **Intuition is our inner Emotional Diplomacy at work.** Let it talk! Listen to it before acting. Think!

The quantum connection of particles that are separated makes one electron feel the movement of the other in the most fantastic and unexplainable way. "***Such quantum connection explains our intuitive feelings about the people that we are emotionally connected to. The powerful links of energy in intuition can even connect us to "the dead that are the transformed form of energy,"****(Dr..Dmitriy Sidorin)* Robots that become more self-conscious now will eventually get the energyof the soul. A humanoid will inherit the mental-emotional structure of the engineer, programming**it**. But humanoids will never have intuition because **it is the divine creation of humans** connectedto the *Universal Field of Information* that requires human receptive perception.

# Digitized Emotional Diplomacy is Our Future Human Courtesy!

# 10. Moral Intelligence Without Negligence!

Emotional intelligence is inseparable with moral intelligence. It is another angle of the aware vision of yourself in the world. Be mindful that we generate it at the lowest level of self-growth-**Physical Intelligence.** **MORAL INTEGRITY** needs to be instilled in us as early as possible, andin many cultures, it is a natural rule of life. With your growing intelligence, **your morality shouldtake an individual form**, not that of the society, a family, a friend, or any common wisdom.

**As the form of your life is changing, the content of it should be altered accordingly!**

The concept of morality is inseparable with ***"pure reason."*** *(Emanuel Kant)* According to Kant'sethical philosophy," ***without insightful reasoning, there is no moral philosophy of pure reason***".The foundation of life is based on "*a working criterion that any rational agent supposedly employsas a guide in making his own choices and judgments." Richard S. Haier*) *Haier* also writes,

**"Intelligence and morality are the result of brain development."**

**A MORAL YOU** is the person with a ***purified mind and a stable, integral morality.*** Gradually,through experience, extensive reading, thinking, and watching the people that are much wiser thanus, we come to realize that our destiny and the whole purpose of being here are guided by the forces that have a much larger perspective on the world stage and that are based on the ***godly principle of goodness versus evilness***. These principles are instilled in us by our parents, and theyare the explicit values of any religion. Thanks to this moral foundation, you can say self- suggestively: ***"I can remove my past emotional knots and deal with the present-day warts."*** Remember the words you use; create the states you have! We all know that ***moral self-perfection***is a life-long commitment that requires a lot of reasoning and will-power, and ***a lot of self-purifying, both mentally and emotionally***. On top of that, one should not fall into being a prude, or a moral phony.

**"Take your broken heart and turn it into Art!"** *(Meryl Streep)*

During intelligence enrichment and spiritual enlightenment, we start merging our consciousness with everything, expanding our intelligence beyond the physical, emotional, and moral intelligence. Through this expansion, ***we are accumulating inner integrity t***hat gets rooted inside.At a certain age, it will ***reach a critical mass.*** It's the state of **PSYCHOLOGICAL CATHARSISthat** we go through when the realization of the true value of things happening to us hits us. ***We start reasoning at a higher level of self-consciousness!*** That's how important our troubles and tribulations are! **Among the most ruining concepts is the money-chasing one.** It pollutes our morality and self-growth and contaminates our souls. So, don't confuse morality with frugality!

# Have a Rebirth of Self-Worth!

# 11. Live and Love by the Code of Love!

We live on the phenomenal planet that is governed by ***the emotional energy of love***, and it is unique in the space which is predominantly mental**.** ***Having gotten the gift of love from berth, weneed to consciously develop it throughout our entire life.*** The breath of love was blown into us by *Jesus Christ* that had planted the seed of love in us, starting ***from the Above.*** According to neuroscience, ***love has a neurological basis***. The scientists prove that when the neurotransmitter of love, called ***dopamine,*** is released in the brain, it contributes to a raise in energy, motivation, and feelings of euphoria or elation. *The Neurological. Love* can be engrave in a humanoid (*See a wonderful movie "Her"****), but the authenticity, sincerity, and the depth of love are our domain!***

**Love is hardwired into the structure of the human brain and the heart!**

From the spiritual point of view, I believe that **love has a structural nature!** It should be grownin the five levels, too - p***hysical, emotional, mental, spiritual, and universal,*** starting with the universal level, in this case. Amazingly, our common cultural observation is "***True love is blessedfrom the Above!*** First love is normally sacred and pure, and therefore, should not be ruined.

Experience teaches us that if love starts *on the **physical level***, it can or cannot ***grow emotionally.*** If it passed this level, it could grow up to ***the emotional / psychological level*** of mutualunderstanding. It becomes much stronger, but it is still not enough for the love emotion to last. ***It needs the mental back-up*** when the two people should connect on the intellectual level. They, *according to Edgar Cayce*, need to have***" the union of purpose"*** in life. If love has sustained thegrowth in these three levels, it is still prone to die because ***there is no spiritual connection*** in it. Therefore, people from different religious backgrounds can hardly keep their love for long.

***The final, universal level of love growth*** connects the exceptionality, the God-granted giftednessof both people, when each partner has a great respect for the innermost goal of the other and is ready to back him / her in its realization. **Only love that has grown through all five levels canlast!** *(Check out the book "The State of Love from the Above!", 2019 or Love Ecology,2021)*

That is why a one-night stand or a love relationship that is based just on **the physical spark** cannotlast long. Like anything else in life, love must grow together with a person's intelligence and consciousness, and it is much better if it gets sparked up at the universal, or the spiritual level first.It then most likely will mature mentally, emotionally, psychologically, and physically. Such love relationship is endless! Most importantly, love needs to be studied and learnt from a very early age and for the rest of our lives. ***Even the love of God is not just granted, it's earned!*** Learn a short prayer of love: "***Dear God, throw me a handful of stars and bless me from the above with love!"*** A pure heart is always blessed with the love from the Above.

# Master the Holistic Culture to Avoid Your Love's Rapture!

# Love Intelligence without Negligence!

*(Best Pictures/ Internet Collection)*

# Moral Negligence Kills Love Intelligence!

## 12. Self-Illumination is in Love and Self-Growth Elation!

To conclude, we all need **LOVE INTELLIGENCE** to drive our emotional mobiles, but **Love Skill must be taught, it is not inherited!** We and our kids are **heart + mind** connected till theirpuberty when socialization gets on the scene, and the opinion of piers, older friends, and siblings predominate. Therefore, "***we need to be especially emotionally refined in a marriage union. "Matrimonial aura is the mental child of both mates.***" (*Laurel Blyth*).

**"It's not the laws of man that wed two people; it's always Go**d!" *(John Baines)*

Love propels us to the place of our destination without any dismay and with a great sense of self-power. ***So, the universal and spiritual magnetism of love is our human stuff,*** and no neuro-linkswill ever be able to imbue the most sincere, deep, and self-sacrificing feeling of love in a humanoid.Their engineered love is meant to just serve us.

Any change translates as stress, but science proves that ***the greatest stress we have is lack of love.***Love -loss or love- betrayal stress generate inner pain, anger, discontent, irritability, and self-guiltin the outcome. Most substantially, these negative emotions ***cloud our life with disbalance,*** insecurity, haste in decision making, and indifference to other people. Command to these negativeemotions consciously and auto-suggestively, **HALT!**

Delete your depressive thoughts, going out into the Sun. Raise your head and look at the clouds where the loved ones that are gone wave you from to up-lift your spirit and teach you to love yourlife in it. When your mood sags, induct yourself with some mind-sets from this book.

***I am my best friend; I am my beginning and my end! / Love is me; love is my philosophy!***

The words ***"I don't care! / What do I care? / I care less!"*** are emotional pollutants for love intelligence. They must be eradicated from your mind and the language. **We must care!** No one can consider anyone a good person if he / she is indifferent to whatever situation he / she might face in life. ***Adjusting to life emotionally, we raise our self-consciousness as human beings!***

*Auto-Induction*: **Love does not adjust to me; I adjust to love!**

Also, let us compliment and complete each other! Add the formula below to you **Auto-SuggestiveMenu,** too. ***Illuminate yourself for the people you love in space and time, and do not whine!*** They are not here forever! Induct the formula below to enhance your luminosity and compassion.Being a luminary is a hard job of Love Intelligence!

## The More Love You Create – the Better is Your Fate!

*(End of Stage Three - Meta-Level, Emotional Intelligence)*

# The Intellectual Odyssey of a Seeking Soul

*Vistas of Intelligence to install next:*

***10. Universal* Intelligence** — **Self-Salvation**

***9. Spiritual Intelligence*** — *( Universal Realm )*

***8. Social Intelligence*** — **Self-Realization**

*7.* ***Cultural Intelligence*** — ***( Spiritual realm)***

*6.* ***Financial Intelligence*** — **Self-Installation**

*5.* ***Professional /Creative/Intelligence*** — *( Mental Realm )*

***4. Psychological Intelligence*** — **Self-Monitoring**

*3.* ***Emotional Intelligence*** — *(Emotional Realm )*

***2. Language Intelligence*** — **Self-Awareness**

***1. General Intelligence*** — *(Physical Realm )*

*/ Physical / Cognitive / Digital /*

**Neuroscience + Digitized Psychology + Self-Awareness =**

**A New human + humanized machine's Psycho-Culture!**

*You can upload any piece of advice with the technological device!*

## To Make Your Life Intellectually Wise,

**Generalize -Analyze - Internalize - Strategize – Actualize!**

# <u>Stage Four</u>

*Meta-Level of Intelligence* - ***Emotional Dimension*** – ***Self-Monitoring***

# Psychological

# <u>Intelligence</u>

*"The unexamined life is not worth living!"* *(Socrates)*

*The Fractals of Intellectually Spiritualized Beings:*

***Form*** + ***Content***

**<u>(Body+ Spirit+ Mind) + (Self-Consciousness + Universal Consciousness)</u>**

= **A New, Intellectually Spiritualized and Digitally Enhanced Being!**

**Inner Dignity is the Base of Our Equanimity!**

# Psychological Intelligence without Negligence.

# I'm Aware of My Joy and Despair!

# 1. Change Your Psycho-Culture!

***Psychological intelligence is the stumbling stone on the path of* SELF-RENAISSANCE.** It is even more complicated than the Emotional Intelligence because it relates to the psyche of a person, a very fragile mechanism of our brain that requires an educated ***psychoanalysis*** (*Sigmund Fraud*).The shortcomings in the psyche of humans and the humanized machines might be eliminated in tandem, too. *David Hanson's company* in Hong Kong is producing wonders in this respect.

> ***"The law of personal survival is executed on the psychic level."*** (*Sigmund Fraud*)

However, ***the basics of the Psychological Intelligence must be acquired by us and instilled intothe humanoids, too.*** The knowledge of the psychological basis is crucial in the education of children. It should later be enriched with the knowledge of the basic structure of the brain and its main functions. In the *Introductory Part* of the book, I have stressed the necessity of changing our**CULTURE OF THINKING** and **FEELING i**nto the holistic one by reshaping our *physical, emotional, mental, spiritual, and universal* frames of thinking that are inseparable with our growing self-consciousness. **Our common ignorance in this field is appalling!**

Working on the psychological intelligence, ***we are sculpturing our own psycho-culture***. Psycho-culture is the culture of **a willful monitoring of our psyche** that we do not have due to the lack ofthe Psychological Intelligence. ***Self-love, will-power, and the sense of Self*** are the most importantaspects of the psycho-culture that we must capitalize on in life." **Who am I?** is the pivotal questionto ask yourself. Next, you need to constantly **fortify your spirit auto-suggestively** with the inspirational, rhyming boosters of your choice They are the shot-cuts to the mind and the gluing element of the spiritual fractal of your life that you should unify consciously at will.

**Form** + **Content**

**(Body+ Spirit+ Mind) + (Self-Consciousness + Universal Consciousness)**

In sum**, *the first goal*** of changing our psycho-culture is the development of the ability to focus onour thoughts and emotions to channel them to the right actions according to ***the Law of the RightHuman Behavior*** *("Right is Might" by R. Wetherill)* **"Think, say and do what is right, and refuse to think, say, and do what is wrong."** As we change our thoughts, we ***change the physiology and neurology*** in the body with the help of special chemicals, called neurotransmitters. The body is full of neuro-logical connections that ***form the pathways*** that are mostly governed bythe unconscious brain. *The cosmic **Law of the Right Human Behavior** and the **Law of Gravity***, channels us along the right paths in life. We should intuit them consciously and ground our negativity with **SELF-GRAVITY** that needs to be installed in humanoids' common cloud as well.

## To Self-Sustain, Put the Mind in Control of the Brain!

# 2. Self-Acceptance is Essential!

***Self-acceptance and self-reflection are inseparable with Self-Management and Self-Refining.*** even though at times, your ***Self-Worth*** might be very low. If you accept yourself the way you are,and admit your inadequacies, you will get more determined to change yourself through self- management. Thus, you'll be better equipped to give yourself the command, **"Halt!"** at the rightmoment. Do not use standardized images of people to boost your self-esteem. **Compare and compete only with yourself!** Perceive, think, and feel in your own unique way! Consciously assess the motivation of other people for using external references to control your behavior and exercise power over you. Listen to your intuition, self-gravity sense, and the inner integrity guard.

**You think what you are, and you are what you think. Put them in sync!**

Just get the implication of these words and be a winner in your mind, not a victim of the hectic life around you. Think differently and respect yourself for the boldness to do it! Have a clear-cutidentity of your personality that has a clear-cut plan of action in the mind.

**In the world of individuals, victimization is impossible!**

In this context, inner strength means being able to stop trying to get everyone think and feel whatyou are thinking and feeling, expecting a valid piece of advice. Learn to be **SELF-SUFFICIENT** and **SELF-RELIENT!** The concept of existential loneliness is ruling our lives now. ***Adjust to it!***

Auto-Suggestive inducting that this book is based on affects your psyche in a very subtle, simple way. So, learn to manage yourself much better because self-suggestibility hits exactly ***theemotional deficiency*** that you might be experiencing at a certain moment, and it adds mental energy to your ***sagging spirit*** at any place and time. You are the only person on Earth who knowswhen that right time strikes. **Help yourself help you!**

*Auto-Induction:* ***The Sun will still shine whether I frown or smile!***

Using **AUTO-HYPNOSIS**, ***you realize the rule of being your own best friend,*** and you synergizeyour inner space, developing a physical habit of perpetual happiness. Self-respect becomes the modus operandi of every action you undertake! Remember that "***one of the basic causes of sickness is unhappiness, and one of the best healers is super joy.***" *(Paul Pearsall)* But feeling super-joy does not mean crazy fun or pretentious, fake and exaggerated excitement.. Unfortunately, many people follow this model of behavior, trying to be" ***cool.***"

*Auto-Induction*: ***I enjoy the inner balance of my Super-Joy!***

# I Know Who I Am and Who I am Not! That's My Fort!

# 3. The Ultimate Result Vision. Use Technology for Self-Ecology!

I have mentioned above the necessity to create ***the Ultimate Result Vision (URV)*** of your **SELF-ACCULTURATION** in the mind as the compass on this rocky path. The main auto-suggestive mind-set (I **can, I want to,** and **I will!**) is very flexible in this respect. You can induct anything, ***making the sub-conscious mind follow the lead of the conscious one***. New pathways will be formed in the brain, or ***the neurology and the physiology*** of your body will work in sync.

***The first goal*** is to. upload the *URV* of **the Vintage You** into your smartphone and have it at handto remind yourself of this mind-set below. Build up a **NEW ME IDENTITY!** Take your attentionfrom the past to the present and the future. ***Make this transformation technologically mutual!***

*Self-Induction:* **I know who I am and Who I'm Not!**

*I am a strong, calm, and determined master of my firm will!*

**I can...; I want to...; and I will...!**

Working at your self-perfection auto-suggestively, you will finally develop ***the holistic connectionbetween the left, concrete, and the right, abstract, hemispheres of your brain.*** We need to put them in synch to gain a better self-control in life. *"The way you think is the way you feel!" Sadhguru)* A happy elation is as short as the strike of the lightening, but self-guilt, depression, andself-pity can last for days.

***The second goal*** of a new psycho-culture relates to is the necessity ***to constantly enrich your URVmentally.*** **As you change your thoughts, you change your psycho-culture!** But your psychological growth must be solely your personal business." *Be the thing in itself!"( Hegel)*Working auto-suggestively at your psychological make-up, you communicate with your unconscious mind by going within and using your imagination and inner vision. ***The unconsciousmind passes the information to the conscious mind.*** Your nerves will set up a magnetic field overthe unconscious brain.

The conscious brain is in turn controlled by ***the higher consciousness*** that shapes the mind of a ***consciously willful person.*** It does not matter in this case whether you are an introvert or an extrovert. Both are great if they are good thinkers. They just conduct their thinking symphonies ina different manner. Thus, you will form ***new neurological connections or pathways*** *that* will raiseyour self-consciousness and make you a much better human being. Scientific open-mindedness is a must, and you need your **PSYCHOLOGICAL AWARENESS** to be constantly enriched with it. ***To be interesting, get interested! Be the best expert of your life.***

**Self-Inducting is, in fact, Conscious Self-Conducting!**

# 4. Psychological Self-Analysis and Psycho-Awareness

Life is a process of challenging us and our responding to it. ***Life does not adjust to us; we adjustto it!*** When the father of psychoanalysis, *Sigmund Fraud*, brought our attention to the importanceof ***psychoanalysis***, he meant that we had to do it ourselves. Psychology is lifeless if we do not impregnate it with self-knowledge and self-analysis in the **Self-Synthesis** - **Self-Analysis** - **Self- Synthesis** way. Psychological awareness is essential for developing any intelligence!

You must reason out your bad and good habits consciously. ***"A habit is something that a body knows better than the mind."*** *(Dr. Dispense) We* suffer because we do not know the essential toolsto operate our psyche with. The holistic **PSYCHO-AWARENESS** maximizes the power of our mind ***in five dimensions*** to rid of lower emotions. I put the spirit in the center of our **FRACTALUNITY** because the spirit keeps the mind and the heart together in us.

**The greatest spirit remains to be the spirit of the unbeatable Thee!**

**Body + Spirit + Mind + Self-Consciousness + Universal Consciousness = A Whole Self!**

To be **PSYCHOLOGICALLY AWARE** means not only being in the know of your ownpsychological keyboard, but also, being **LIFE-AWARE,** or self-conscious every living moment.

*Self-Induction:* ***My aware attention is never in retention!***

Stay aware while driving, talking to someone, walking, working, and exercising. When working out in the gym,***focus on your muscles, or areas of the body that you need to shape***, do not becomean automatic object of exercising unconsciously, as a robot on a remote control. *John Baines writes,* **"The Center of Gravity is in the psychological awareness of a person"**

Being at the end of the psychological tether is too dangerous, and there are too many bad stories testifying to people's helplessness in dire circumstances. It's easy to say, ***"Snap out of it!"*** The question remains **How?** So, the essential knowledge in psychology is a must for every human being, ***and it is most crucial for teaching our children its basics of it at school.*** We need to do itto get rid of the bulling problem that is looming over weak, unsteady, ***uncultured souls of our kids***. **Psychology is the area where knowledge is true power!** The knowledge of our psychological make–up is power only when it is put to action. This action must be taken in traininghumanoids, too. And it should be applied ***in five dimensions of life.*** What we call our **COMFORTZONE** is, in fact, the lack of true self-awareness. To feel psychologically stable, we need to leavethe habitual comfort zone and create psychological **SELF-GRAVITY** of life awareness.

***We need to know how to handle ourselves to be able to handle others!***

## Don't Fight to be Always Right. To Be Right, Be Psychologically Bright!

## 5. "Be Immune to the Poison of Life!" *(Carl Yung)*

To be psychologically intelligent means also ***taking responsibility for your responses to the present moment,*** *with*out resorting for an immediate help of a psychologist or a psychiatrist "***Living in the present, you become the present of God!"*** *(Neil Donald Walsch)*.

**Being psychologically aware, you do not life-spare!**

Life is too stressful for us now to have no idea how to help ourselves in any problematic situationand at any time. Getting an appointment to seek immediate help might be a problem, but life cannotbe stopped. Whatever the problem might, **be your own best friend in it**. Be aware, be in control! **Do not fear, steer!** The best and the most precious thing in the world is ***to be free of otherpeople's negative influences,*** or free from what *Carl Yung* calls "***the poison of life***" that builds upinhibitions or complexes in our psychological make-up

.**"Real freedom is the freedom of the spirit!"** *(Dr. Mather Luther King)*

We need to learn to be in control of ourselves, for we can give the world the best we have only if we have accumulated the values that can be beneficial for it. In this respect, we need to find the way to incorporate both the Western and the Eastern psychologies into our mental framework. ***TheHolistic psycho-awareness demands knowledge about both systems.*** We know that the Western psychology is based on outside, objective observation of people, whereas *the Eastern psychology* is based on ***profound, meditative introspection of self.*** Hence, we should acquire better social skills and get new leads in life *(See Cultural and Social Intelligences below)*. Both philosophies are the indispensable basis for widening our outlook and must be studied at school.

***Then, we will be able to de-magnetize, or to deactivate our conditioned and limited psychologicalpatterns of thinking and feeling.*** We must remove the mental blocks that we had accumulated forcenturies under the pressure of the ***"collective unconscious"*** *(Carl Yung)* that has turned into the "**robotic unconscious. SELF-MONITOR** your life. Be consciously alive!

Finally, ***psychology should be renamed into*** **"JOY-OLOGY"** *(Dr. Paul Pearsall)* the author of the phenomenal book" ***Super-Joy***". that has practically saved my self-worth, smeared by immigration. I saw that book, lying on the pavement of one of the streets of Manhattan on the third day of my arrival in the USA. It was sold by a street book seller to me for just a dollar, but itwas worth a million for me ever since. *Dr. Pearsall* writes that instead of focusing on the abnormalities of human psyche and behavior, the psychological science should instill in us **the joyof living** that will teach us" ***to cure the normal***" and take a risk of being " ***super-normal."*** Have the guts to say to yourself and others:

# My New Phycology is "Joy-ology" + Self-Ecology!

# 6. Demagnetization of the Spirit

It is an accepted fact that it is easy to say," ***Stay away from the poison of life!"*** than to do it. To be successful in this very challenging goal, we must first learn to resist the temptation and then ***to***magnetize our inner core with new values.

*Auto-induction:* **Resist, reject, and reform to obtain a new life uniform!**

We need "***to reverse the triple 666***, signifying the **mind, heart, and sex** into ***the triple 999*"**, or inthe celestial terms, change the evil, badly willed side of us into the divine one that is still there, butgot weak for whatever reason "*Most of us spend our lives bouncing back and forth between the good and the bad sides of life." (Ravi. P. S. Berg).* ***"Be activated by the internal influences for the external responses"*** Don't be a mere echo of the instinct reaction of the majority or, of the "collective unconscious." *(Carl Yung)* Play a solo role in that concert.

***Be unique, personable, and charismatic! Be a New You! Be better than You!***

***Launch yourself every working day*** by heightening the voltage of your spirit self-suggestively! Giving yourself the command a, by counting backwards:

*Self-Induction:* *Five - Four-Three -Two - One* – ***Start!***

***Resisting, rejecting, and reforming are very self-forming!***

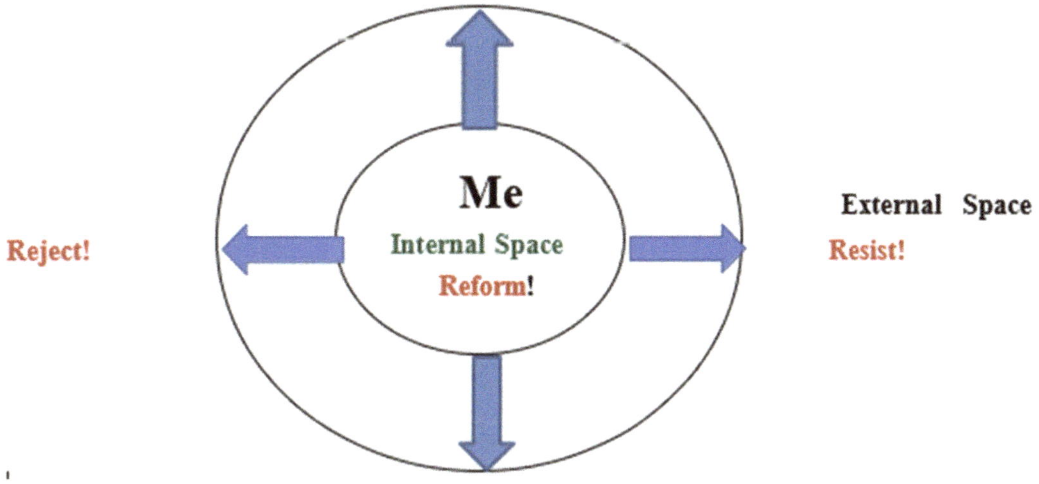


## Expand Your Inner Circle of Resistance to the Poison of Life and Thrive!

# 7. Constantly Reform Your Inner Deform!

Never stop strengthening your ***intellectual culture linguistically, emotionally, professionally, culturally, and socially.*** Self-creation is an integral process that goes holistically in every level. The necessity to train ourselves to respond to life mentally as opposed to reacting emotionally getsfixed into our ***virtual matrix*** now only if we are aware of the importance of shaping our minds.

Auto- Induction: ***I must be reasonable and bold to fit into a new life's mold!***

We all make mistakes and have sins of wrong thinking, saying, and doing. All a person needs is torealize what he had done wrong and not get into the vortex of more wrong doings by force of habit.

**The inner matrix should be personalized, not digitized and robotized!**

It appears that only with the belief in God and conscious **rationalization of thinking*;*** governed by the Emotional Intelligence can we resist, reject, and reform our imperfect inner uniform. The words of a great British writer *Somerset Maugham* at the end of his wonderful novel *"The* ***Moon and Sixpence"*** are very convincing in terms of the punishment we get if we have no immunity tothe poison of life.

***"The mills of God grind slowly, but they grind exceed small!"***

There is another point to consider. We cannot be weak for a long time. The people that get discouraged or depressed and those that make drugs or alcohol an obsession, justifying themselvesthat they are too stressed out , must remember that .they are inevitably sinking into the **PSYCHO-HOLES** that they have dug for themselves but have no inner strength to get out of. It happensbecause they do not blame themselves for the weaknesses that have demagnetized their inner magnetic core and depleted the mind of the mental energy. *Dr. Fred Bell* writes in this respect,

***"General demagnetization of a spirit always starts and finishes with the brain."***

Therefore, it is ***vital*** **to INSPIRE** ***yourself*** and others with **LOVE FOR LIFE,** compassion, understanding, mercy, self- belief, and the incentive for self- actualization in life. Don't let yourselffall into the depression stall. I totally agree with *Antony Robins* who declares

***"Depression is a self-inflicted disease!".***

*(See the books on Inspirational Psychology for Self-Ecology- "My Solar System", "It Too ShallPass!" and "Self-Taming!)*

**"Drink the Juice of Life, but do not Drown in it"** ***(Sadhguru)***

# 8. Introspection is Self-Inspection

***Emotional intelligence and psychological intelligence are inseparable in our mind.*** The difference is only in our awareness of the emotional life we have on the one hand, and of the peculiarities of our psyche, on the other. **Introspection is an indispensable part of intelligence development.** The challenge of life is always there, but to sustain the pressure of a fast-developingeconomy and our hectic life, we need ***to take an observer position*** on the intellectual, emotional, professional, and moral levels of our personal growth.

*Auto Induction:* **If I don't discipline my thoughts, I live backwards!**

We are either aware or unaware of our psychological states. When we are in the dark, we are proneto panic, to seek someone to share our worries with. In the long run, we develop more complexesthat do need professional help to be removed.

***When we are psychologically aware,*** we know what to expect of ourselves in different situation and how to avoid unfavorable consequences. The vital thing is ***to make the right choices*** in life by giving yourself time to think of the consequences of your actions first and consider which sideof the equation (*light or darkness*) this choice is on "***You can be light, you can be darkness, butyou cannot be both!***"*(Eleanor Roosevelt)* Get used to X-raying your thoughts, words, and actions. Introspection of the past actions triggers the body's self-healing abilities and boosts self-esteem. It ***is not good to be totally dependent on a psychologist or a psychiatrist*** to explain to you what's going on inside you, unless, of course, you absolutely need it.

**Don't ever whine. Shine!**

You are the only person in the world aware of it, and if you start looking at yourself ***retrospectivelyand introspectively***, armed with the essential knowledge about your psychological make-up, you will be your first urgent help. ***Do not pass judgment on yourself or anyone's cell!***

Help yourself at the first signs of anxiety, depressive thoughts, or irritability. *Try to illuminate thecauses for such feelings and give yourself an inspirational boost.* You've made the right choice!

***He who holds up his head will always go ahead!***

*Basic Self-Induction:*

**I Am My Best Friend.**

**I Am My Beginning and My End!**

# 9. Psychological Adaptability

Overseeing oneself psychologically is essential in self- actualization, but it is not enough. We aresocial beings, and our psychological wellness is related to the to the wellness of the society we livein. The **PSYCHOLOGICAL ADAPTABILITY** reflects our conscious behavior in the society.

***Never have relationships been so problematic!*** Your success on the path of self –creation will depend on how well you handle yourself and others. To begin with, we need to stop judging andgive others more psychological air to be able to self-actualize, too.

## Shed off the burden of judgment of yourself and others!

We need to accept the fact that relationships are not meant to make us happy, ***they are the tests ofour self-consciousness*** that grows with each new relationship, causing us to become more conscious, more compassionate, considerate, and loving. It needs reasoning, rationality, and emotional intelligence, to begin with.

***Making the other wrong is easy; making the other happy is a challenge!***

Recent trends including high-tech communication, overseas travel, global economy and a dramaticrise in immigration throughout the world have created new challenges for our **psychological adaptability** that is, in fact, our psychological awareness. If you observe unconscious behavior inyour partner, stay aware of it yourself, so your knowing could keep you from blowing!

**Be Consciously Conscious!**

***Always know exactly what you are thinking, feeling, saying, and doing!***

Know if you react to someone's unconscious behavior, you become unconscious yourself. Conscious attitude leads you to responding, not reacting, to understanding, rather than being understood first, to giving rather than taking, to loving rather than fighting. Psychologicalintelligence becomes increasingly essential because ***our interaction face -to- face and mind-to- mind depends on it***. When you face a conflicting situation, remind yourself:

*Auto- Induction:* ***I perceive with my soul to understand and to console!***

Psychology teaches us that "***perception is more important than intent***". In other words, what weperceive in a conversation with someone is much more significant than the intention of the personspeaking to us. Here goes the rule of the *Language Intelligence* again:

## Say What You Mean and Mean What You Say!

# 10. Self-Management is Learnable!

Self-management is our priority business, or it's the private business that we need to run to our own advantage. Self- management is a process of trial and error like anything else in life. have onthe windshield of your car the mind-set:

*Self- Induction:* ***I'm driving through the time and space with a smile on my interface!***

Self- hypnosis helps a lot to connect with your sub-conscious mind. Later, this work will put youinto ***an altered state of consciousness*** that will prompt to you the right mind-sets and channel youto the right actions. Simple reading of the mind-sets that rhyme and repeating them to yourselves,or even learning some of them by heart can do the trick. The right mind-set will just pop up in the brain at the right time to back you up, if need be.

*Auto-Induction*: ***My mental wings blow off the emotional swings.***

**Mobilize Yourself!** You might not realize it, but your personal capabilities will be getting more powerful without making others less so. See yourself as very ***charismatic and magnetic!***

**Charisma is an excellent case of self- management!**

Unfortunately, comparison dictates us most frustrations and judgments. ***Compare yourself to the best of you not anyone else!!*** Stop worrying whether you fit on the comparison scale with your successful friend, an older brother/ sister, or a neighbor. You do not have to look outside yourselffor self-assessment. Say to yourself auto-suggestively, ***"I know who I am!"***

*Auto-induction:* ***I do not compete or compare. I am just the best here and there!***

Once you grasp the power of self-induction, charisma can never be taken away from you. As a matter of fact, anyone can become charismatic with exceptional knowledge and a bunch of competences / intelligences that we focus on in this book. The best doctor is our cheerfulness ***because it's a healthy spirit that gives us energy, happiness, and makes us irresistible.***

People gravitate to the one with a strong personal magnetism that radiates light and a positive attitude. Drive your incredible personal mobile and be proud of it while it is driving Your self- inducting, or the words you use create the states in you.

**Inspiring words-inspire; despairing words-despair.**

By managing the mind, you manage yourself, and you channel your life to **SELF- DISCIPLINED** or **SELF-MONITORED** happiness consciously. *As Reid Hoffman says,*

## "Manage Yourself like a Well-Run Private Business!"

# 11. Every Human Contact is a Responsibility!

In sum, much too often we hear, "***I didn't mean that***!"- "But you said it so, and I got it that way due to the wording of your thoughts!" In view of this situation, the monitoring of our thoughts insync with the words in which we frame them is vital for an adequate communication. Our ***new behavioral skills require our mandatory ensuring of a smooth and effective relationship*** with our family members, our friends, the people we date digitally or in person, our colleagues, and ouraging parents. We need to work out **the adaptability pact** with everyone we care for.

Auto- Induction: ***Do not be accusatory; be "amusatory!"***

Get into a habit of X-raying the holistic make-up of a person you deal with on a regular basis. Be more perceptive. Do it with aware attention, creating the other person's profile integrally, thatis *in the physical, emotional, mental, spiritual, and universal* planes. **Synergize your interaction** with everybody by establishing the connection in the heart and the mind ***The choice of the One and the Only should be holistically and consciously monitored***, too. The **HEART + MIND SYNERGY** is the guarantee of the well-made choice.!

Cooperation creates the energy of understanding, compassion, and love in a co-creative fashion. Compassion and understanding are of paramount value here. You will be surprised with your ownability to manage the feelings of other people in the way that helps you both. It's like playing a different musical instrument, being proficient on the main thing - ***the notes-reading in the soul ofanother person.*** Conduct the inner orchestra in yourself with aware attention and life-affection!

Adaptability should also be our main behavior pattern with our children. It is vital in your relationships and in your resisting the habits of discriminating people by skin color, the religious affiliation, or just lack of education that they never had a chance to get. people are amazingly interesting if you give yourself time and patience to explore their brains and souls.

**Acceptance and adaptability** are the **LIFE SKILLS** that you need to keep developing for the entire life Try to reform only your own deform. ***Live and let live!*** Be a self-guru, then people willgravitate more to you! There is no one to fix ***your inner gravity*** that holds you together in any situation. You do not need any un-related advice that you are seeking from your friends, loved ones, etc. ***You are the only one who knows what to magnetize your soul space with*** and how to keep it from getting it partially or totally de-magnetized. Read King Solomon's proverbs by the number of the day you need advice. It is a great **SELF-GRAVITATIONAL** device.

# Inspiration or Desperation – That is the Life's Equation!

*(End of Stage Four – Psychological Intelligence / Soul-Refining, Emotional Dimension )*

# The Intellectual Odyssey of a Seeking Soul

***Vistas of Intelligence to Install Next:***

**The greatest Art of all is to Self-Install!**

*10. Universal* Intelligence — **Self-Salvation**

*9. Spiritual Intelligence* — *( Universal Realm )*

*8. Social* Intelligence — **Self-Realization**

*7.* ***Cultural Intelligence*** — ***( Spiritual realm)***

*6.* ***Financial Intelligence*** — **Self-Installation**

*5.* ***Professional*** */Creative/* ***Intelligence*** — *( Mental Realm )*

***4.Psychological Intelligence*** — **Self-Monitoring**

*3.* ***Emotional Intelligence*** — *(Emotional Realm )*

***2. Language Intelligence*** — **Self-Awareness**

***1.General Intelligence*** — *(Physical Realm )*

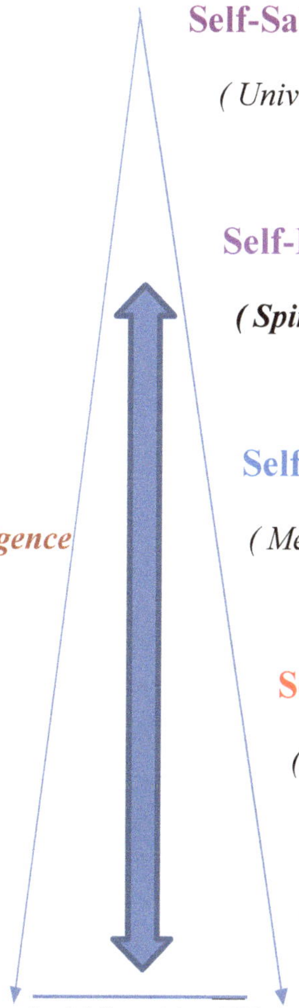

*/ Physical / Cognitive / Digital /*

*(Upload your smartphone with a new Psychologically Enhanced tone! )*

***Self-Installation is the path to Self-Salvation!***

**To Make Your Life Intellectually Wise,**

**Generalize -Analyze - Internalize - Strategize – Actualize!**

# Stage Five

*Mezzo-Level - Mental Dimension –* ***Self-Installation***

# Professional

# Intelligence

***Self-Installation – Division 1***

***Inflate Your Personal Pride with the Professional Might!***

**"Try? There is No Try, only Do and Not Do!"**

*(Steve Jobs)*

## Professional Intelligence without Negligence!

## Procrastination is a Silent Killer!

# 1. Professional Self-Installation

There is nothing compared to the feeling of a single accomplishment on the path of self-evolution professionally. ***You attain the superpower over yourself,*** and you feel that people start gravitatingto you in hope to gain some inner strength that you emanate and the peculiar knowledge that you have. So, true professional intelligence goes beyond one's specialty and professional credentials. Because only a real professional can connect **THE FORM AND THE CONTENT** of the productof his mind or put the professional knowledge and skills in synch.

**Professional intelligence is the unity of the knowledge and skills.**

***It's your personal, unique know-how!*** Over the years, a true professional accumulates very special, selected knowledge, **INDIVIDUALIZED KNOWLEDGE**. You develop *exceptional professional skills* that no one can take from you because ***the Metrix,*** the complete framework ofthe thought-process of your is beyond anyone's reach. Every person's ingenuity is amazing!

**Everyone has a unique structure of thought!**

You can be motivated by ***Steve Jobs, Elon Musk***, ***Jeff Bezos*** and other extraordinary minds, but you cannot duplicate the structure of their being *physically, emotionally, mentally, spiritually, anduniversally.* ***Everyone's thinking architecture is unique.*** That is why, even in the cases of the **intellectual theft**, the intellectual property cannot become totally plagiarized. Any idea that you might have voiced out cannot be picked up by somebody's brains in its holistic entirety. You canbe robbed of your idea partially, but it will never be realized in the way you see it yourself, ***in its unique wholeness.*** So, do not limit yourself with the fear of plagiarism. **What's yours is limitless!**

*Self-induction:* ***I don't want to be ever told about my imperfect professional mold!***

A professional mind is an endless source of ideas when it keeps operating with confidence of its uniqueness. **PROFESSIONAL CONFIDENCE** penetrates the essence of other disciplines and gets to the hidden causes of all that is manifested as an effect in your own area of expertise that remains to be unique. Such situation proves Norbert Winner's brilliant observation of today's professional world when we need to send a person that is seeking the solution of a professional problem "*to the next door down the corridor*", instead of answering his question then and there. All a real professional need is ***a holistic visualization of the problem and its immediate X-Rayingin five dimensions for the solution***. As I keep reminding you, self-growth should be channeled holistically, following the paradigm below. Only then can you **OUT-BEST YOURSELF**! Competition with Self only enhances the professional zest and boosts creativity.

**Generalize - Select- Actualize!/ Synthesis - Analysis-Synthesis!**

## 2. Professional Awareness

I have devoted so much attention to the ***psychological intelligence*** above because it is essential indeveloping the p*rofessional intelligence* or building up professional *self-installation in life.**Professional intelligence is the domain of the mental dimension of self-growth***, and the greatestpart of our life is spent in the professional environment and its psychologically healthy / unhealthy climate. You might be in the wrong field professionally to for just material reasons but ***try to fit your exceptionality*** in any field of knowledge because your eventual search for the right field of self-expression will only be enriched with **holistic awareness**.

**Being the best is a tough test!**

**PROFESSIONAL EDUCATION** must **be** taking a person beyond the limits of the old, generallyaccepted knowledge, ***monitoring the brain strategically and much more proficiently*** with the helpof technology. A future professional must be motivated not with just getting a certain number of credits but with acquiring the knowledge that goes beyond the limits of a certain program and enhances **PROFESSIONAL SELF-CREATION,** not to get stagnated in ***stereotyped thinking.***

You have already molded somewhat your physical, emotional, and psychological intelligences; and you are more able to monitor your language and speech. At the next level of our intellectual evolution, you need to live by the professional code that every one of us needs to work out for himself. And that always depends on the ***rightly chased professional field.*** that is the prerogativefor self-realization in life. You must **INSTALL YOURSELF** professionally and realize yourselfon the path of self-perfection, proving your God-given exceptionality.

***Professional intelligence*** is also a life-long commitment, the same as self-development. Recognition should be given again to thc book "*The Start-up of You*" by *Reid Hoffman and Ben Concha* that view professional development in a very actionable way. Obviously, professional intelligence is not a stagnant phenomenon! The authors, for instance, suggest most rightfully "***developing more transferable skills.***" They name the skills of high value – "***speaking / writingskills, general management experience, technical and computer skills, people smarts, and international experience, or cultural skills.***" Self-development will help you preserve ***"the* AUTONOMY of THINKING*"*** and go forward, no matter what

Also, "***youth does not have a monopoly for professional intelligence, age does***.*"(Reid Hoffman)*With age, we sharpen our professional skills and intelligence, but we also can fall into the trap of**stereotyped professional thinking** that is often in the way of new ideas and a fresh vision. Mediocrity is another pest on this way. Unfortunately, it's true that ***the leadership usually belongsto the mediocre***, but you should not get discouraged by that if you are professionally aware.

# "Income Rarely exceeds Personal Development!"

## 3. Self-Genesis Strategically and Tactically

Acquiring the ability for ***reasoning and rationalizing life*** is a great skill that, unfortunately, fails us sometimes because we are emotional beings first, at least at this point of our human evolution.The pendulum of the uncorrected life mistakes often crushes and destroys our creation of Self. Obviously, we must respect our feelings, too, and be aware of their harmful or beneficial effect onour personal mind-set. Self-awareness is being instilled in humanoids now, but the level of **objective reflectiveness and intuition** are unsurmountable for them. ***Intuition is more emotionalthan mental, while conscience is more mental than emotional.***

Therefore, we always have intuitive feelings, warning us against wrong choices in life that we haveto reason out later thanks to ***the twinges of conscience*** that prick us inevitably. Therefore, every mistake we have is a lesson to learn, and there is always another lesson inside each lesson.

*Auto- Induction:* **The choices we make dictate the life we live!**

The point is to start seeing with the heart and stop repeating the same mistakes repeatedly. But intuition will prompt the right action only if we rationalize its message in the outcome first. In other words, the paradigm of rationalization of life demands that we first think of the consequencesof our actions strategically, then take the action and watch ourselves ***performing it tactically.*** Thus,we'll be paying conscious attention to the causes of your actions, not only their consequences thatis ***we synchronize our mind and the heart.*** You must not be impersonal!

*Auto- Induction:* **I am learning the art of seeing with my heart!**

As an egg-headed intellectual, you are gradually shaping yourself into an egg-form, into a cocoonthat can transform into a butterfly. This is the process of accumulation of the magnetic power of the mind, on the one hand, and the healing power of the emotions, on the other.

*Self-Induction:* ***I am what I think and feel, but I sometimes need self-consciousness refill!***

That is why we look for solitude to let our mental and emotional wounds heal and to give ourselvestime and space to rationalize the life steps that we take. To accomplish that, it's paramount to always have the initial, holistic vision of the **ULTIMATE RESULT VISION of SELF- INSTALLATION** in the mind. It's your reference point that will help you compare the URV ofone time-space period to another, obtaining a qualitatively much better vision of self in space andtime. If every misstep in life is a message to learn; we need time and space to process that lesson consciously through our **CONSCIENCE GRID** ***first*** strategically, then tactically.

# Thus, you are Molding Yourself over and over into a Whole You.

# 4. Rational Thinking and Intuition Development

***Rationalization starts with changing the way we process the information*** that invades your brainnon-stop, and rational thinking is at the core of any professional intelligence. Only conscious thought process can channel ***cognition to action,*** when each bit of information is filtered by the brain through the existing knowledge, gets selected or sorted out for its validity, and is later ***organized into the chunks of information*** on the basis of a conceptual principle and in the holisticway, presented above. (*Mental Architecture, Chunk 9)*

Start with reading or listening to the piece of information that you need ***analyze quickly for its validity*** and ***select the information t***hat is of interest for you. Be professionally aware and wise- *inthe physical, emotional, mental, spiritual, and universal dimensions.*

## Generalize -Analyze - Internalize - Strategize – Actualize!

Next, you need to **conceptualize it** into the messages that you get, reading between the lines ofthe text, a verbal or written one. Thus, you will be able to channel your thinking consciously andrationally in the cause-effect framework much better because you will feel more responsible for every thought you have, every word you say, and every action you take. Never view the problem in one dimension only! Aware attention should be given to the problem at every level - *physical, emotional, mental, spiritual, and universal.* Only holistically reviewed business has the perspective to grow***! The Artificial General Intelligence*** pushes us to the vision and a quick solution of any problem holistically**. Super intelligence is in the highest professional need!**

The reasoned out and structurally governed thinking should always be rational and conscious, andit needs to be monitored by ***the Universal Law of Cause and Effect***. It should not be affected by the patters of thinking that are already established in the brain and that ***generate compulsive thinking*** and therefore, an action impulse that we might feel sorry for later. A person's consciousmind should always be in charge to block ***the greatest professional enemy*** **–** the **INERTIA OF THINKING.** Therefore, **self-education must be a never-ending process of self-growth.**

*Richard W. Wetherill* calls it "***a professional wart*** or ***distorted thinking****" (The Tower of Babel")* It is very common among people now, led by the automatic patterns of behavior more than ever. Automatized thinking is the result of an unconscious, automatized, and patterned thought.

Other than that," ***distorted thinking causes emotional energy to be used destructively and the result of distorted thinking is emotional and physical debility that gradually becomes complete***."

***As it is Above, so, it is below!*** *(The Hermetic maxima)*

## As you Think, So, You Are!

## 5. Holistically Think to Put the Mind and the Heart in Sync!

Science proves that most people ***over-analyze the information.*** The left hemisphere of the brainis responsible for this process. We first need to deal with critical items of information as fast as possible, ***conceptualizing them in the cause-effect fashion.*** Next, at this new stage of our evolutionary development, generated by the digital revolution it is paramount for us to put our reasoning in sync with feeling. The **HEART+ MIND** connection is essential not only in our job relationships, but it is crucial for our creative abilities that are based on this unity. To put it differently, we need to process the in-put information though our individual mental-emotional gridsynchronically, forming new patterns of conscious, rationalized behavior. The goal of this boo is:

: **Intellectualize your heart and emotionalize the mind! Be One of a kind!**

Putting the mind and the heart in synch, ***we are processing the informational in-put through thetest of consciousness and intuition,*** the act that will always point us to the right behavior solution-making, or action-taking. Neither just the mind, nor the heart will do the job on their own.

**Our intuition is, in fact, our conscience!**

***The conscience and intuition are inseparable entities***. They are testing every thought we have, every uncontrolled word we say, and every automatic action we take. What we call "**a guilt –trip**"that is always crashing us after any wrong-doing is our conscience at work.

To keep self-unspotted from the world, we need to tune to our conscience that in the form of intuition works as **the *scanning mechanism*** of the cause-effect framework of the mind-heart unanimity. Intuition is the basis for your communication with the Universal Intelligence – God! Intuition is, no doubt, the propelling mechanism of our self-development, too, because it is connected to conscience that is the testing mechanism of our growing consciousness, governed byour tuning to the station "God" for our intuitive self-assessment and guidance in our very turbulentlife.

*Self-Induction*: ***I tune myself to the station God for my inner peace reward!***

It's noteworthy to mention here that the ability to unify the mind and the heart is the brain operationthat builds up **the mental-emotional grid** as the Fort of our inner gravity and personal magnetism,and most importantly is channeling us to rationalized behavior. The mind must be in charge, silencing the chatter box of the brain. Put the strain on your brain with the mind-set:

# I Make My Heart Smart, and the Mind Kind!

# I'm One of a Kind!

# 6. Professional Consciousness and Conscience

***Holistically developed professionals are the cornerstone of today's economy***, and professional intelligence is at the forefront of evolution. We need to raise the level of professional awareness to a higher **PROFESSIONAL CONSCIOUSNESS**. It is the most paramount way to raise the economic productivity and the quality of the economic output in a business, as well as the way tomake our businesses more viable.

Professional consciousness is something that often sags due to the desired security of a job that isnever there and loss of intellectual or financial motivation that damages professional conscience of a person. **Professional consciousness requires constructive or proactive thinking** to be donein terms of the quality of the work that you perform daily, and it should not be damaged by. the colleagues of less motivation and less intellectually – up-dated. Such thinking has to be always logically-structured and strategically-disciplined on the basis of a person's unique know-how thatwill remain in constant and flexible growth, seeking perfection till the end of a person's life, addinga lot to his / her sense of self–realization in life.

. ***There is no success in self-quest without a person's professional self-actualization!***

In fact, professional intelligence is at first governed by the **PROGRAMMED INTELLIGENCE**that any person gets with the college education on the road of career searching. The one, who has***processed information in his own unique mind consciously,*** is capable to handle it skillfully in the future. Unfortunately, there are still cases when the professional intelligence of people that areheading a business, a company, or a governmental institution leaves much to be desired and, therefore, it becomes the obstacle for many advanced minds on the way of the realization of theirwonderful ideas. You need spunk to overcome these obstacles.

***"Character, not circumstances, makes the man!*** "*(Booker T. Washington)*

In sum, professional consciousness is, by and large, a display a strategically structured, professionally disciplined, insightfully deep, and very creative thinking.

It is, in fact, very proactive thinking because a highly professional person should be able to generate new context, concepts, and objectives in the most constructive way, putting both the heartand the mind into his job. ***No doubt, a career choice that goes with the calling of a heart is a must!*** Therefore, we should help young minds discover their exceptionality as early as possible.

***Practice ⟹ professional awareness ⟹ professional competence ⟹ professional integrity ⟹ professional conscience!***

# Professional Conscience is Mind + Heart Competence!

# 7. Beware - Stereotyped Thinking!

***To get to a new dimension of professional consciousness***, we also need to operate any problem in the most *conscious way* because the professional world is also being screwed up by the automatic, thoughtless job functioning. **Brain-wracking is an exercise for purposeful thinking** in the search of the solution of any problem.

**"*Put your mind on fire; be proactive!*"** *(Reid Hoffman /"The Start-up of You")*

Unfortunately, once a person gets stuck in a framework of his professional obligations, he stops caring for the quality of the job done and often functions as **a working robot**. The obtained academic degrees and tenues are often in the way of a person's further professional and scientificgrowth. The bureaucratic policies and instruction kill the initiative and the creative urge of a worker. It happens because we generate certain **STEREOTYPED MENTAL HABITS** at work that condition our thinking unless we are fully aware of such a possibility and do not ***let the mindget inflated with worked-out patterns of thinking.***

***Professional thinking is inseparable with constant growth in professional awareness***.

The use of the old patterns in professional problem-solving generates **PATTERNED THINKING**that is a dead end in the professional sense. Not to fall into the trap of patterned thinking, *Reid Hoffman* suggests**,"** ***You should never leave a start-up's mind position*"** to always challenge yourself with the unpredictability and insecurity at work. He writes that" the rules of the labor market have been permanently altered. **"Ready, aim, fire"** has been replaced by" ***Aim, fire, aim,fire, aim, fire!"***

What a great way to put it! That is so true because only a start-up professional is always agile, thinking-wise. The Law of the Right Human Behavior comes to our rescue here, too. *"**Being rightis the identification with a mental position. There is nothing that strengthens our professional skills more than being right. However, this skill needs work, too.**" (Richard. Wetherill).*

Therefore, we have more managers now than real thinkers or true leaders that are able to reboot their minds with new thinking patterns, devoid of programmed, overworked ones. We can get ridof this harmful state of thinking if we develop ***the aware attention skills*** **holistically**, in every intelligence and in five main dimensions of self-growth.

We can acquire these skills if we consciously process any information that we get through a ***previously sorted out and well–organized knowledge grid***. It is possible only with the holistic vision of the reality that demands a flexible and constantly renewed approach to a job...

## "You Need to Forever Be a Start-Up!" *(Reid Hoffman)*

# 8. Inflation of Professional Thinking!

We live in a technologically sophisticated universe. However, since we delegate more and moreprocessing to computers, perceptual thinking remains to be often forgotten, whereas it is essentiallyhuman. Only human interaction helps generate neurons within the cerebral cortex and form new memories. ***In fact, forming new connections in the brain, we essentially rewrite our consciousness*!**

In-coming information gets organized into patterns, and the mind follows the patterns automatically, following the recipe. However, the best recipe can be screwed up due to the **inflated thinking** of an employee who works on the automatic pilot, on the one hand, and is driven by thefeelings of insecurity of losing his job, on the other.

***Inflation of thinking*** generates fear of action and hinders the mind in the process of decision making or creation of a new thinking vibration. The mind of an employee gets emotionally and mentally inflated by the negative vibration. Science proves that our thoughts, emotions, desires orimpulses are only vibratory states. In our personal or professional life, ***everything revolves aroundthe vibration***. If we are happy, we generate positive vibration, or we feel happy due to positive vibration. Sadness, fear, discontent, suppressed anger generate negative vibration. Each person has a mental wavelength of his own, according to the vibratory frequencies of his intelligence.

***Our thought makes us intelligent according to our mental vibration.***

The fear of being politically incorrect and breaking the protocol generates negative vibration, andas a result, inflated thinking. That is not to say that following the job instructions is bad. While organization and discipline are essential at work, they should not be in the way of a creative mind-flow of the employees and their ***freedom of self-expression.*** **A healthy synergistic professionalclimate is essential here.** It stimulates employees to apply creative thinking, and eventually, everyone benefits. **Mind + mind** and **heart + heart** links must be established in sync.

Next, we might have to battle **the inertia of thinking and intellectual apathy** that mostly make superficiality a bad habit Not to aggravate the situation, impedes the search for the critical information in every intelligence realm. It might be needed to resolve a professional problem effectively. Being **HOLISTICALLY AWARE** will spare you much trouble.

To conclude, ***learn to work with people below you***. But dealing with the people below, don't getaffected by the phenomenon of **PROFESSIONAL INFLATION** due to the prestige of your position. Remember that nothing is permanent under the sky, and you might end up where you hadstarted. **Personal vanity is our worst enemy here.**

# Make Self-Worth your Main Boss!

## 9. Professional Superficiality

There is another very significant point to make here. ***Science proves that most of our thinking isdone in the perception area.*** Perception is how we look at things., and how aware of life we are. The neurological understanding of the brain affects how we live our lives. Your professional intelligence absolutely must be enriched with the new insights from Neuroscience about the flexibility of the brain and its plasticity. This knowledge is essential for the Super Artificial Intelligence and the creation of the **self-aware humanoids.** The development of these innate qualities in our brains should be t***he main goal of our technologically baked up education!*** Suchnecessity applies to any field of knowledge and any expertise as the demand of our times.

**Knowledge about the brain has huge cognitive benefits professionally.**

We alter synaptic connections in the brain and change the language behavior patterns, thus changing a regular human interaction **to mind-to-mind communication**, as well as making the interaction with a machine become more and more meaningful in the virtual reality. **Professionalsuperficiality** that is often the result of an inflated professional thinking is impossible in our interaction with an exact machine. So, the digitized reality demands a more **INSIGHTFUL PERCEPTION** of our job duties that the computerization of economy is instilling in us. Also, being right does not mean necessarily that we immediately must find someone who is wrong. "***Thefinger pointing to the moon is not the moon*". *(Buddha)*** Being right means acting with aware attention, applying proactive thinking, purposeful thinking, or the cause-effect thinking that are resisting patterned thinking in its core. ***New professional awareness demands conscious recognition of patterned thinking.***

One of the reasons we often must deal with ignorance at work lies in the fact that ignorant peoplelack intelligence due to their **PATTENED THINKING BLINDNESS** that is **ALWAYS STEREOTYPED**, never questioned, challenged, or destroyed with a new flood of consciously processed professional information that demands not only ***a mind-to-mind interaction, but also aheart-to-heart connection between the members of the team.*** Our evolutionary development depends on the unity of our hearts and minds that are now in defiance!

Developing aware attention is another way to battle the inflation of thinking. Aware attention is an electronic commodity now that helps to confront this situation. Aware attention is the ability ***todiscriminate the knowable from the unknown*** or from the patterned application of knowledge that is killing many businesses now worldwide. To channel attention of the website visitors in theright way guarantees success only if you manage to hook their aware attention and thus establish **a mind-to-mind contact** with them.

# Conscious Professional Thinking is a Must here!

## 10. New Professional Holistic Methodology

Our_managerial culture often lacks professional intelligence, and it so happens that a manager sometimes cannot explain the job to the employees in a simple, aware way.'

We have a lot of managers everywhere and **very few real leaders** that are able not only to decidesmartly and hit more lucrative targets, but they can also change the framework of business, science,and life overall. Senior management often cultivates ***the culture of fear*** when people are afraid toexpress their thoughts which could have changed the texture of any professional enterprise. We should feel free to demonstrate our professional competence and express it freely without any red-tape restrictive instructions or the fear to lose a job. ***New professional methodology demands inclusive leadership and holistic professionalism!*** It means that professional assessment should be done on five levels, holistically, as well as the training for a job

**Mini, Meta, Mezzo, Macro,** and **Super!**

Mini level – *A* ***start-up position; very little knowledge, limited job skills; no experience***

Meta level – ***Knowledge and skills are still disconnected, some experience is acquired***

Mezzo Level – ***Knowledge and skill are synchronized, a person has worked his own know-how, professional competence, full awareness, professional intelligence.***

Macro level - ***Great professional intelligence, spontaneity and creation, leadership skills.***

Super level - ***Professional consciousness, great cultural and social skills, enrichment in otherexpertise, holistic professional outlook, synergistic professional skills***

**Professional competence has a new meaning now** because our professional expectations have changed drastically. The new time demands that holistically educated and self-created individualsperform in the best way possible in the multi-national, globalized, electronically governed world.

In sum, the ability for innovative and constant growth in every profession is the given now. Professional growth has to come in new ways, and our mission is to make it inclusive of the latestscientific developments, technological advancements, and the methods of working together synergistically with diverse teams of people.

***A competent representative of any company should be a representative of theholistic intelligence and competence inclusively.***

## Personal Magnetism is Not What you Do. It is How you Are!

# 11. Professional Self-Assessment

Professional self-reflection is the guarantee that you will never fall behind in your knowledge andskills. Constant work at both is a necessity because **a real professional has to have them in synch,**especially during our digitally changeable times.

***Knowledge + skills = the flow of professional life!***

Listed below are fifteen guidelines that can be used for any professional self-assessment, after youhave scanned yourself for your knowledge and skills levels presented above. Of course, every fieldof knowledge has its peculiarities, but there are some features in every profession that are of a general value. You can make your own list of guidelines, depending on the expertise you have andthe job that you are practicing. Quiz yourself to build more professional confidence and fortify your professionalism with more awareness.

***"Excessive Ambition leads to Disaster!"*** *(William Shakespeare)*

## Self-Assessment Professionally:

Guideline 1 – **Assess your job responsibilities.** What does it take to do this job superbly? Comprise a strategic plan of action. The one who plans better accomplishes more. Studies reveal that planning is a cognitive task. It demands the development of analytical and conceptual thinking.De-automatize your work. Assess the necessity to fill your mind with new knowledge and skills. *(Cognitive Intelligence, Chunks7-12)*

***No pseudo-learning! Develop professional awareness non-stop!***

Guideline 2 – **Assess your job performance.** Face any challenges objectively and consciously.

***When the going gets tough, the tough gets going!***

Guideline 3 – **Professionalize your speaking habits. Talk the professional talk**, use the professional vocabulary. Manage you bad language behavior patterns. **First think then speak!**

***Don't verbalize everything that comes to your mind!***

Guideline 4 – **Get the feedback from others.** Studies show that people get to know us better in reflection. They reflect us and thanks to their input, we get valuable professional feedback

Guideline 5 – **Build up on your natural talents.** See what makes you special and unique and how you can incorporate those talents into your job.

Guideline 6 – **Seek out excellence in all that you do,** all that you say, and all that you are. *"Youshould not be just the best; you must be the single one!" (Steve Jobs)*

**Guideline 7 – Continue to invest in your mind. It's your biggest asset**. Professional mistakesare the result of undisciplined, negligent mind and lack of professional awareness.

**Guideline 6 – To be interesting, get interested!** Go with the science flow! Be well- informed1

**Guideline 7 Live by the professional moral code!** - You need to trust your moral barometers,be in charge of your moral behavior at work as much as you try to display it at home.

**Guideline 8 – Learn to synchronize intelligence and professionalism!** Connect them by theformula: **Synthesis – analysis – synthesis** when resolving any problem.

**Guideline 9 – Keep professional integrity**. Don't undermine it for anyone's sake. Always build **a "shorthand' profile** of the person facing you. Trust your own intuition more than anyone's.

**Guideline 10 – Synergize the working place, radiating the win-win attitude.** Enlighten theprofessional space! Apply" **network intelligence"** skills.

**Guideline 11 – Always think of the outcome that you will accomplish in qualitative terms**.Neutralize an opponent's intention for control with your intelligence, then take power!

**Guidelines 12 – Don't ever surrender to ignorance**. Resist it and fight it with the power of yourprofessional logic and informed attitude**.** Growth vs. Stagnation!

**Guideline 13 – Give yourself and everyone around an emotional and professional pump up.** Learn to compliment and praise any one even for a smallest accomplishment.

**Guideline 14 – "Make a leap from potentiality to actuality!"** Acting on the mind-set auto-suggestively uploaded in the brain is the best way to upgrade your professional toning.

**Guideline 15 – Most importantly, don't knock you down trying to compete with others.** Build yourself up competing only with yourself!

**The Fractal of Intellectually Spiritualized Being:**

***Form*** + ***Content***

**(Body+ Spirit+ Mind) + (Self-Consciousness + Universal Consciousness)**

**= An Intellectually Spiritualized and professionally more advanced You!**

# I am a Unique Cell; I Keep on Surpassing Myself!

*(End of Stage Five / Division 1 / Professional Intelligence)*

# The Intellectual Odyssey of a Seeking Soul

***Vistas of Intelligence to Install Next:***

**Fertilize your thinking with the creative God-guided inkling!**

| | |
|---|---|
| *10. Universal Intelligence*<br>*9. Spiritual Intelligence* | **Self-Salvation**<br>*( Universal Realm )* |
| *8. Social Intelligence*<br>*7. **Cultural** Intelligence* | **Self-Realization**<br>*( Spiritual realm)* |
| *6.**Financial Intelligence***<br>*5. **Professional /Creative/Intelligence*** | **Self-Installation**<br>***( Mental Realm )*** |
| ***4. Psychological Intelligence***<br>*3. **Emotional Intelligence*** | **Self-Monitoring**<br>*(Emotional Realm )* |
| ***2. Language Intelligence***<br>***1.General Intelligence*** | **Self-Awareness**<br>*(Physical Realm )* |

*/ Physical / Cognitive / Digital /*

***"It's Not enough to be the best. Be the Only!"*** *(Steve Jobs)*

## To Make Your Life Intellectually Wise,

**Generalize -Analyze - Internalize - Strategize – Actualize!**

# Stage Five, Division 2

*Step Five* - ***Professional Intelligence*** - ***Self-Installation***

# *Creative*

# *Intelligence*

## *Without*

# *Negligence!*

*Inflate Your Personal Pride with the Creative Might!*

**"To Be a Creator in Hell,**

**You Must Believe in God and Yourself!"** ***(Ben-Groin)***

# Exceptionality Beats the Reality!

**“Creativity is the Ability to Think Abstractly”**

*(Carl Yung,” Creative Imagination “)*

# 1. Your Creative Vigor is at the Trigger!

The will to create is encoded in our DNA! We live at the age of AI, an incredible imagination andsimulation technology. Its mind-boggling creativity is changing our world exponentially every day. I love this country for the level of ingenuity around me. Th***e Creative Intelligence*** that had created us initially is remaking us all the time, and it is holistic in its essence. So, let's give creativity an open air because creativity is everywhere!

**Creativity is the pursuit of collective wisdom!**

***Creativity is a burning sensation*** to let the world know about your idea and find the way to presentit and get recognized. ***It is the main asset of any good professionals and the goal of his / her life.***Creative expansion of the mind occurs when you listen to your inner voice and trust your intuition,or the hunches of the right brain, while following the logic and the structural lead of the left one. There is no true professional intelligence without the upbeat contribution to it on the part of the creative mind of its owner. But how can we know something without feeling something?

**RATIO + EMOTIO = CREATIVITY!**

For 28 years that I have been in this country, I keep marveling at the American creativity and ingenuity. I love this country of doers and creators, of advanced minds, of intellectual and spiritualgurus. A deep well of brain potential resides within us all, but only those that keep developing their creative intelligence "***find the best way to tap into the inner savant'*** *(Scientific American", August 2014)* **Creativity has no race, no skin color, and no politics.** It's universal, and it needsto be recognized and supported because a creative person is driven by a very altruistic urge to better the world. ***"Creativity is the ingenuity in action."*** *("Scientific American Mind", July 2018).*

Creative skills let the scientists explore undiscovered artistic capabilities of the right brain, the brain that is putting everything together against all odds. *A **plane gets off the ground against thewind!*** In 2008, a cognitive psychologist ***John Koumiss*** and his colleagues discovered that the brain activity of people, who used insights, differed significantly from those who preferred just analysis. ***Elon Musk is the best example of a real creator*** that the humanity needs now.

**A real creator does not just talk about the future. He makes it!**

***Carson's reports*** from highly creative individuals suggest that to prompt your brain to create, youshould first "***wander the world with an open mind***!" To develop the right hemisphere of the brainthat is responsible for our unique creativity, listen to classical music, beautify y our speech skills,and synergize your life and the lives of other people with the **mind+ heart** unity, The real doers of life have exceptional **SELF-WORTH!**

# Creativity is Me; Creativity is My Identity!

# 2. Change the Form and the Content of Your Thinking!

To be creative means questioning the existing **form ± content** of your life and, most importantly,putting them in sync! ***"For knowledge to be useful, it must be creatively applied."*** *(Albert Einstein)* The moments of insight*," the Aha moments"* occur when the brain spontaneously reinterprets information "*to reach novel, nonobvious conclusion*" *(Nessa Victoria Bryce) As* I haveindicated in the part "*Cognitive Intelligence,*"

**Creativity is the sync of the form and the content of the object of creation.**

Interestingly, *Steve Jobs*, a genius of our age, managed to change and to connect the ***form and content*** in all his electronic marvels in the unity of an unbelievable precision. Our present-day most advanced giants of technology, amazing entrepreneurs and philanthropists *Elon Musk, Bill Gates, Jeff Bezos, Mark Zukerman, Reid Hoffman,* and many others are moving the technologicalworld forward unbelievably. It takes a true visionary to bridge the visual design and its inner capability together or think holistically and strategically. ***The desire to pursue the self-quest andself-creation is driving only the creative minds!*** Emotion is also the bridge between conscious and unconscious minds. ***Creative emotion is called inspiration!*** So, the power of creativity needsto be directed willfully and patiently. Otherwise, it disturbs the creative inspirational fractal that agifted person is always forming holistically.

***Form*** + ***Content***

**(Body+ Spirit+ Mind) + (Self-Consciousness + Universal Consciousness)**

That is why creative people are always very vulnerable and too sensitive when the unity of their creation is disturbed with ignorance or mis-consideration. They are conscious of any hostile attitude to the product of their creation. Therefore, they are tenaciously fighting for it with those who are unconsciously or deliberately hampering the way of their creation. ***The untrained intellecthas no power over a genius mind!*** Life abounds in the examples of such battles of the minds.

Next, anyone can enhance his creative thinking ***by expanding the knowledge base,*** by learning new skills, and by talking to someone outside his immediate professional network. We boost our creative potential through expanding a general outlook and a spontaneous thought. ***Creative intelligence develops a person's mind more than any other intelligence***, and it is uplifting the mind to the **Living Intelligence** level because a creative person gets out of the loop of the "***immediate gratification***" of his wants that demagnetize his ***Eureka moments*** and low his self- consciousness. He is driven by the inner determination: **I can…. I want to….and I will…!**

# To Be the Best is a Tough Test!

# 3. Actionable Imagination

Creative intelligence is never ordinary because **it repels the patterned, inflated thinking.** *( See Cognitive Intelligence above)* Science proves that most creative thinking is done ***in the perceptionarea of the brain*** that determines how we get to look at things. Creative imagination is the intelligence that goes beyond the limited understanding of things. ***Creativity is becoming more and more a collective phenomenon now*** thanks to the digital revolution because it is more than an expression of individuality**.** It is a team thinking and feeling. **It is intuition driven!**

Thus, creativity is taking shape in a social context, but it is always channel led by a leader's mind,**the one that has the holistic vision of the common creation!** Actionable imagination does not come out of nowhere. Since we delegate more and more processing to computers, **perceptual thinking** remains to be the most important to preserve because it ***organizes the in-coming information into patterns.*** The mind follows the patterns automatically. The purpose of perceptionis to allow patterns to form first and then to use them. ***A creative mind focuses on the proximity of form and content of these patterns in a holistic framework.***

**Selection + Organization + Strategizing =A Creative Product**

Creative intelligence requires conscious recognition of the patterns, not to be automatically drivento a pre-programmed solution. It rebels against the old patterns and creates new connections between the neurons in the brain that get programmed for the impossible, unexpected, and unique.We are all familiar with the beautiful words of *Albert Einstein*, "***Imagination is more important than knowledge. Knowledge is limited, imagination is infinite!"***

**ACTIONABLE IMAGINATION** is the fuel for the mind, and creativity is the process of proactive thinking and feeling**.** Wonders of engineering, technology, medicine, space exploration,you name it, are countless. The world gets much of its multi-national "***ingenuity striptease***" fromthe Artificial Intelligence breeze. Halleluiah to the incredible hard work and perseverance of the most advanced scientists, researchers, designers, and engineers. They are our global pride!

***The American Ingenuity Award*** was given to ***Elon Musk***, a great innovator who is rocking the world now with his innovations and most sparkling intelligence. **"*When I am concentrating on something, I tune everything else out. I am curious about anything."*** *(Elon Musk),* Mr. Musk islaunching a private cargo into space, designing upscale electric cars, and changing the way we think about the solar power! He is also a self-taught engineer that explores the frontiers, so daunting and so dangerous! He wants to defeat the incumbent old thinking in rocket building andfundamentally change the way we travel. As the *CNN's Amar C. Bakshi said, "**His career encapsulates a brand of big, big thinking,***"inspiring young minds to follow him, declaring,

## In My Creative Quest, I Manifest My Best!

# 4. Making Things Possible is Not Impossible!

Opponents of Elon *Musk's phenomenal contribution to the world of innovations* tell us that he hadthe money to do that. Unquestionably, that's obviously true, but ***he made the money building upon its potential power.*** He, obviously, has the will power to expand his intelligence, his limitlesscreativity, the most surprising awareness in many areas of science, business, and technology.

***What a wonderful example of self-propelling and most creative intelligence!***

Positive emotions, a romantic love, the desire for self-realization and self-establishment act as boosters for creative intelligence. Aha moments come while we are observing nature, reasoning on the problems, scanning books for interesting information, or just meditating.

***We question the right, creative hemisphere*** that harbors our intuitive insights, and we get the unexpected answers to our questions. Just imagine **the creative audacity of thinking** that ***Elon Musk*** and ***David Hanson*** with their very creative teams are displayed to the world!

**Only the minds that question the impossible make things possible!**

But this magic happens only ***in the intellectually loaded brains***, the ones that are in constant betamode. "*Such thinking can be called enlightened for it is based on awareness that enlightens the mind and redesigns it. Such designing goes beyond the boundaries of the mental dimension up to the spiritual one.*" (*Osho, "Intelligence*") ***That is why creative people are loners.***

They fly in their minds, they are out there, somewhere, "*not quite together*" in the perception of ordinary people. We must be particularly grateful to such people and help them in every way possible. Their sensitive human brains add immeasurably to the richness of life. They have madetheir choice, and it always justifies itself because **they have created themselves!**

**(Body+ Spirit+ Mind) + (Self-Consciousness + Universal Consciousness)**

**Before you can get something, you must be somebody!**

Finally, the management culture that we have touched upon in the previous section, has to be builton creative intelligence of the best leaders, those who are ***able to strategize life, rather than focuson immediate tactics in the resolution of any problem.*** The people that create humanized robotsmust be real leaders because they instill their personal characteristics into the neuron network of the humanoids. **Be such a leader!**

## "It's Not Enough to Be the Best, Be the Only!"

*(Steve Jobs)*

# 5. To Be Interesting, Get Interested!

. Finally, the long-awaited times for ***the intelligence and advanced creativity*** have deservedly come to the forefront of our economic and social development. Creative intelligence generates ***thenew light of thought*** at a peculiar wavelength concentrated only on one thing. A creative personalways has ***a skill mix and the mix of intelligences*** that help him illuminate the patterns of thinkingin his head. His knowledge acquisition is in constant flux.

**Without up-dated awareness, knowledge is dead, and creativity is impossible!**

Any creative person should be familiar with the names of Arthur ***Clarke, Robert Heinlein, Ray Bradburn, Isaac Asimov, bothers Strugatsky***, and any other science fiction writers that had predicted the wonders of the present world. Aware attention to the potential marvels of their discoveries has become a real business commodity now.

People that can think beyond the professional frame-work are in demand now The required peculiar mix of professional skills is always based on the information that has been accumulated by the inquisitive minds thanks to the aware attention that they pay to a much wider range of expertise than their own, *on the one hand* , and thanks to the ability to question the ultimate resolution of any problem ,*on the other.*

***Creative intelligence is ruling the world of inadequacies. That's why we are progressing!***

Creative people are always independent thinking. It is vital for those that do the hiring ***to tune to the creative vibes*** of such jobseekers and be **HOLISTICALLY AWARE** themselves, not to fail to place them at the right time. These must be the most intelligent people in any company. Naturally, ***the development of the professionally aware attention is indispensable*** in any creativeprocess**.** The code for understanding and communicating between aware people is very special. Itis being developed in social networking on the Internet. But to connect to the ideas that *the Universal Intelligence* is communicating to us *in **a mind-to-mind fashion***, we need to get rid of **the mental paralysis** of the fear to lose a job, self-doubt, inferiority, and inadequacy that are reining in many departments. The professional and human self-worth are s the main force here.

In sum, we need ***to tune our*** **AUTO-ANTENNAS** (*our intuition*) to the Universal Mind during our inner active / passive meditation and act confidently and "doggedly" toward success! If you are *self-reliant and can meditatively attune yourself to the Universal Mind,* you will be successfulwithout any doubt. Just be patient. The Universal mind has you scheduled.

**Competence generates creative confidence!**

**Vision + Idea = Creation + Determination+ Success!**

# 6. Doubt is an Immobility of Thinking!

On the path of your ***most important mission in life***, there should be no doubt in you because doubtpoisons all endeavors and generates the **IMMOBILITY OF THINKING** that is ***the malady of people in trouble***. Doubt is, in fact, part of life if we view life as the field of energy that is fluctuating in a negative / positive fashion non-stop, exactly as our personal lives do. Doubt is alsogenerated by some one's reaction to our actions. People rain on our parade and change the fluctuations of anticipated positive result in our minds if we let them do it., having no ***SELF- WORTH*** umbrella over the head. *(See the book "Self-Worth")*

Quoting *Hegel*, let me say again, ***"Be a thing in itself!"*** The greatest German philosopher pointsto the representation of the ***content-form*** phenomenon. ***"If you view self-formation as a life phenomenon, the inseparable unity of your form and content will be pointing to the necessity ofdeveloping intelligence in synch with the will, as only such oneness completes a person as a subject of the world."****(Hegel)* Re-charge your aware attention by processing your idea through the holistic paradigm of **synthesis- analysis -synthesis** or

**Internalize – Select – Externalize.**

**Strategize - Optimize - Actualize!**

Developing the ***"thing-in-itself"*** wholeness will make you self-reliant and self–sufficient. It is theprerogative of a talented person to choose "***a narrow way***" in life. *(The Bible) Robert Schuller* says, ***"Don't be simple-hearted about your success in front of people, only in front of God!"*** Only with your inner wholeness and optimism will your mind and you psyche be enthused. You will become the happiest man / woman on earth because ***you are blessed with a mission*** that keepsyou going on and on, without being hampered by anyone's judgment or skepticism.

. **An optimist lives in a self-created bliss because his world is never amiss!**

***Great minds teach us to clear the residual emotions and see the roots of our failure.*** Then we need to tap into the energy of true knowledge and the latest information in science that feeds the brain and charges its three-fold transformational capacities:

**1. Find Yourself! → 2. Motivate Yourself! → 3. Commit Yourself!**

According to Ben–Groin, "***To be a creator in hell, you must believe in God and yourself!"***
*Put this self-induction on the windshield of your car:*

***I am driving through the time and space with God on my Interface!***

**In My Creative Life Quest, I Commit to Be My Very Best!**

## 7. Steer the Self-Judgment of Your Idea!

A true creator judges himself the validity of his idea, without paying attention to the critics and allthose who rain on his parade. ***Elon Musk,*** having started his new company ***SpaceX,*** had a lot of"well-wishers" whose predictions he had discarded , having launched his own rockets into space.That's the best example for every forward-thinking mind must act.

**Some more recommendations that you might want to consider.**

*1.* ***Process every new idea on your own; do the self-judgment of its uniqueness.*** An identity decalis not a reward for being unique, but it's the inner assuredness of being individually unique and having the ability to stand up for your uniqueness, like ***Jesus, Socrates, Gandhi, Mother Teresa,Mather Luther King, Junior, Steve Jobs,*** and many other most advanced minds.

***2. Assess the depth and clarity of your idea and its potential novelty and advantages with awareattention.*** *Train humanoids, if you are in that business, to be sensitive and personable like yourself. Help the cloud that programs them collectively be installed with the best of our qualities.*

***3. Examine the key ingredients of your idea holistically -*** in the *physical, emotional, mental, spiritual, and universal realms.* **Have the robots screened in these five levels, too**.

***4. Be -conscious every living moment! You know who you are and who you are not!. Work atboth!***

***5. Anticipate the costs and the result. Keep rechecking them and transforming them.***

***6. Share your ideas only with your co-thinkers. Create an aura of thinking together!***

**"We are mating through creating, and we are creating though mating!**

In sum, we all have within us untapped potential and undamaged areas that lead to new open doors.***This potential needs to be actualized in the most personable way***. Reading avidly and expandingyour general outlook, admiring art, and listening to classic music feeds actionable imagination.

**"What you are is much more powerful than what you will be!"** *(N. Goddard)*

No wonder, Einstein was playing the violin to boost his actionable imagination that was ruling his" ***thinking laboratory.***" Creative intelligence is the intelligence that sees everything as a problem that can be fixed without any excuses, but with a lot of determination, zest, and love.

## A Creative Being does a lot of Intellectual Brewing – Creating, Failing, and Redoing!

# 8. Professionally Advanced Thinking You!

Your new ideas may seem bold and unrealistic, but you know better. You are the one who is on permanent communication with the Universal Intelligence. It is guarding you in theimplementation of your crazy ideas and it pushes you to defy **the gravity of a common thought.*"For an idea that does not at first seem insane, there is no hope."*** *(E. Einstein)*

Most importantly, don't knock yourself out trying to compete with others. Instill in your mind themind-set: **In my professional cell, I compete only with myself!** Making yourself better and betterwith each new working day, you never stop visualizing the expected result, not just in the materialized form, but in self-realization form. True personal growth normally occurs during the professional life of a person, especially when this person puts ***both his mind and the*** heart into his
/ her job, loves it, enjoys it, and grows with it physically, emotionally, mentally spiritually, and universally in terms of realizing the **RESPONSIBILITY OF ONENESS** ***with*** the business of his
/ her life and the people that he / she shares his /her life with.

This attitude should be applied to the interaction with humanoids that are becoming more and more***organic and artificial, or self-awareness trained***. Let the right action become your *physical, emotional, mental, spiritual, and universal might* of the mind able to go beyond the terrestrial boundaries and become super-human. Be proud of yourself! Your vision will inspire the mind. ***Your connection to the Universal Mind will back you up!*** The result will manifest itself withoutfail! ***The Law of Attraction works without a fraction!*** Remember, you are the Only One on Earthwho has the idea that is propelling you to the stars

*Auto-induction* **In my mind, I'm one of a kind!**

***There wasn't, there isn't, there won't ever be any one like me!***

Self-hypnosis will give a boost to your creative powers.

***I know who I am!***

***I am a calm, strong, and determined owner of my firm will!***

**I can. I want to! and I will!**

, and it'll back up your confidence that might sag sometimes. To boost your creative intelligence,just remind yourself as often as possible of the ***intellectually spiritualized fractal*** of your life.:

**(Body+ Spirit+ Mind) + (Self-Consciousness + Universal Consciousness)**

**= An Intellectually Spiritualized and Professionally Advanced You!**

# 9. Oy Vey! Victory is on the Way!

Psychology teaches us that "***there are no hopeless situations; there are people that take hopelessattitudes.***" In the world of non-stop creation, failures are the best exercise for the brain in this endeavor. Failures are our indicators of the short-cuts to self- awareness and self-confidence. Unfortunately, when we fail on our trial-and-error path, we allow others to drain us of our enthusiasm and the intuitive feeling that the right decision is just around the corner.

***" Beware of lack of perception of your control dramas.***" (*Eckhart Tolle)*

Success-failure are natural fluctuations of a creative intelligence. However, failures happen muchmore often on this path, and a creative mind must sustain them against all odds. The world historyis full of such stories, and they fill us up with awe at the perseverance of accomplished people.

**"Feed your faith in success and doubts will starve to death!"**

**You must always control our attitude.** The degree of your healing after a failure depends entirelyupon your ***ability to restore the inner light of your self-confidence*** and the desire to better the situation with your new special in-put into it. With any negative act we are putting ourselves at risk because we change the charge of the informational field around the body, and it gets disconnected with the Informational Field of the Earth and the universe. ***Disconnection is death!***

***Without a*** conscious control of this connection line and tuning y on our **AUTO-ANTENNA** to the Universal Mind, you are doomed to walk all your life on the cyclical treadmill of success – failure. As a famous business philosopher *Jim Rohm* puts it,

*"**If you must be addicted to something, make it an addiction to winning!**"*

When you failed to do something, the first question should be not **where** the mistake lies, but **why**the mistake occurred in the first place. It is at your fingertips to regenerate the action that had failedand channel your thinking toward success in it. Such attitude speaks of the **change of professionalconsciousness.** ***"If you think you'll lose, you are lost!***" *(Napoleon Hill)*

**"Failures are the best lessons".** *(Steve Jobs)*

***Winston Churchill*** had to start from scratch in his political endeavors four times, and each time he taught himself to take every failure with more calmness and acceptance, generating more confidence and faith in God and himself thanks to his being calm. He kept saying to himself,

## "Don't be Vexed, be Humbled. Victory is on the Way." Don't Spirit Sway!

# 10. Self-Hypnosis is at Hand; Remember About That!

In conclusion, there are always obstacles on the way of our polishing and selling the pearls of ideasthat come as a revelation. We all know how hard it is to fight with ignorance on this path, and howmuch self-confidence, belief, perseverance, and just strong will-power are needed to promote ourideas, especially when we are aware of the fact that nothing is impossible.

***"There is no idea so good that it can't be ruined by a few well-placed idiots"*** *(Scott Adams)*

Since no one is aware what ideas, you are nursing in your wonderful head, you are the only one who can help those ideas mature with the boost of your self-confidence and a firm will-power toback them up. So, keep saying to yourself auto-suggestively the mind-sets that will help you become ***a stoic in the fight with ignorance.*** Every idea is propelled by your self-confidence, faith, imagination, and your uniqueness. **The Air is Full of Ideas for the Connected Mind!**

Self-Induction*:* ***I'm a stoic, not for fun; I get through life with an up-lifted thumb!***

Doing suggestive meditation as often as you manage, you will be able to relax and clean up your head. *(Chunks 12, 13, 14, Physical Intelligence*) Never stop boosting up your willpower with a ***regular reminder that will always help you make better decisions***. Engrave into your memory:

***I know who I am!***

***I am a strong, brave, calm, and determined owner of my firm will!***

***I can…, I want to…, and I will…!***

You can always come up with a better way of doing anything if you inspire yourself on a regular basis auto-suggestively. All you need is to imagine the impossible and put a lot of effort into making it possible first in your inner vision. ***Never lose faith in your mind!*** If the robots can have60 different human expressions, our own mindset must be always focused on assuredness and thebelief in God and oneself*!* ***Our unique humanness must remain unbeatable for the humanoids.***

Often, all that's required is simple activity, instead of passivity. Choosing not to speak up or do anything in the face of obvious impossibilities fades your talents and deletes the possibilities. ***Helpyour professional intelligence infiltrate your job environment***. Turn on your mind and enthuse yourself with confidence. *Napoleon Hill* said, "*When I am right, nobody remembers, when I am wrong, nobody forgets.* **Your creativity expands your consciousness, and that's what counts!"***Continue "being the thing in itself! (Hegel)*

# Preserve Your Uniqueness, Not Bleakness!

## 11. Self-Affirming is Self-Forming!

When time allows and you want to give yourself a boost, do some self–suggestive work in terms of self-affirming, saying to yourself some compliments that are presented below. ***You can never overdo with self-inspiration, for it is never enough!*** Do it consciously and with much ***aware attention*** applied to breathing in the first part of the induction and breathing out the second one..

### Feel your unique presence on Earth! Be Your own Boss!

***1. I'm never in a robot-like trance; I'm governed by my aware attention dance.***

***2. I get lighter in my mind! I am getting refined!***

***3. I discipline my thoughts, words, emotions, and actions without any fractions!***

***4. I am physically and mentally fit; I am upbeat!***

***6. I love, and I am loved in return! I respect, and I am respected!***

***7. I am developing a synergetic win-win mentality; I turn hostility into humility!***

***8. I accumulate a solid emotional account; I raise my words, not my voice!***

***9. I do not listen with pseudo-attention; I am in my aware dimension!***

***10. I listen to be listened to, and I hear to be heard!***

***11. I compliment others sincerely on their success to an excess!***

***12. I build myself up by competing only with myself in every cell!***

***13. My financial intelligence knows no negligence!***

***14. I am the clone of the best of me, see!***

***15. I am happy, no matter what! Happiness is my full-time job!***

### My Life Elation is in Self-Installation!

# In My Mind, I am One of a Kind. I do Not Whine; I Shine!

*(End of the Creative Intelligence /Division 2. Professional Intelligence*

# The Intellectual Odyssey of a Seeking Soul

***Vistas of Intelligence to Install Next:***

**"Money is not the root of all evil. Lack of money is!"**

*10. Universal Intelligence* — **Self-Salvation**

*9. Spiritual Intelligence* — *( Universal Realm )*

*8. Social Intelligence* — **Self-Realization**

*7.* ***Cultural Intelligence*** — ***( Spiritual realm)***

*6.* ***Financial Intelligence*** — **Self-Installation**

*5.* ***Professional / Creative/Intelligence*** — *( Mental Realm )*

***4 Psychological Intelligence*** — **Self-Monitoring**

*3.* ***Emotional Intelligence*** — *(Emotional Realm )*

***2. Language Intelligence*** — **Self-Awareness**

***1.General Intelligence*** — *(Physical Realm )*

*/ Physical / Cognitive / Digital /*

**Money needs brains, discipline, and character gains!**

*(Do not forget to upload your smart phone with a new Self-Exceptional tone!)*

## To Make Your Life Financially Wise,

## Generalize -Analyze - Internalize - Strategize – Actualize!

# Stage Six

*Mezzo-Level* / ***Mental Dimension*** - ***Self-Installation***

# Financial

# Intelligence

***"Money is not the root of all evil; lack of money is!"***

*(Dale Carnegie)*

**"A Man Needs Money to Never Think about It."**

*(Brothers Strugatsky and A Tarkovsky / "Solyaris" )*

# Learn Financial Intelligence without Negligence!

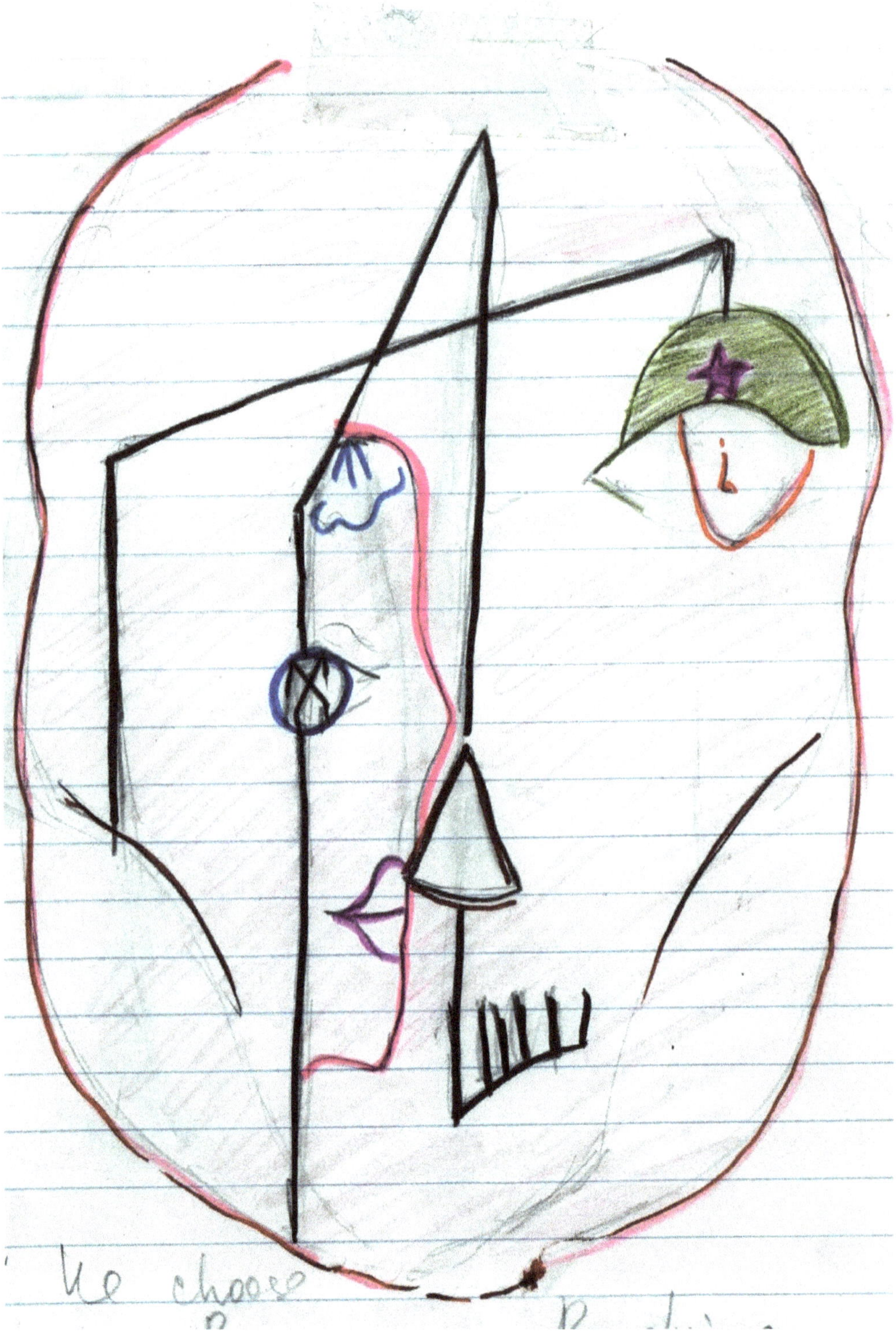

**A New Me is Financially Free!**

# 1. Brain Veins Need Money Gains!

***Financial intelligence is an integral part of a personal growth and self-actualization.*** In my Mission Statement, I talk about the change in the paradigm of our self-development, accentuatingthe priority of ***being over having*** in self- actualization.

**"Money never exceeds personal growth!"** *(Ray Hoffman)*

Money is an especially important aspect in self-realization, and money managing is the skill that is essential in the evolutionary process of an individual growth. However, money is the tool thata self-developing person needs ***to operate consciously*** to base his self-propelling on. So, the title above can be used both ways. Innovations need brains and money investments. The **FINANCIALINTELLIGENCE** of those who have it helps them attain economic independence, lack of whichhinders the process of self-creation in its core.

**To be on the wheels, we all need Money Management Skills.**

**Financial Self-Installation** must be pursued at every level of the holistic paradigm. In other words, we must have financial intelligence to be able ***to digest complicated financial information***and teach ourselves and our kids to manage money, make investments, and harness financial appetites. The earlier we learn to manage our money, the better! Everyone needs to start his self-growth with gaining **FINANCIAL AWARENESS,** preserving his **FINANCIAL INTEGRITY,**protecting his financial privacy. We all deserve to be well-off, wealthy, prosperous, rich, and affluent. These words, though they are synonyms, mean various levels of financial prosperity.

***Financial maturity as any other intellectual maturity needs to grow holistically, or in five dimensions of life, too.*** It means that financial intelligence requires***financial-awareness*** skills andat least general knowledge of money -managing. ***Economics*** must be taught in schools to sow financial intelligence in kids.

True financial competence, ascertained in terms of business or personal needs at each level, will channel the mind in the **mini, meta, mezzo, macro, and super levels,** *or* in a holistic way. Youwill be processing **MONEY STRATEGIZING SKILLS** through the grid of the ***physical, emotional, mental, spiritual, and universal*** dimensions, not just in an in-coming and out-going way. The strategic plan of action that is holistically amplified is the Universal Mind-- fortified!

I have been managing my money through these dimensions for years, and I have always put bothends together. I am a scholar, not a money-maker, but I managed to live at peace with my money,having followed the holistic paradigm all my life, holistically strategizing its coming and going. Ifmoney is dealt with chaotically, it is corrupting people in the bud.-

## Strategizing – Optimizing – Actualizing= Money-Wising!

# 2. Money-Managing is Learnable!

Many financially intelligent people teach us how to manage our money and manage our banking accounts better. We need to study their wisdom and never stop working on our money managementskills. "***They teach us to make money work for us, not us work for the money!***" The advice of ***Suzi Orman*** is so valuable because there are no gimmicks behind her financial solutions. I also recommend some very special books: "*The Richest Man in Babylon*" by George Cayson, the life-changing books by *Robert Kiyosaki and his books, co-authored with Donald Trump.*

**To climb up the Financial Hills, develop your Money Management Skills!**

All these authors suggest many actionable solutions, and, in my understanding, they follow the holistic paradigm of S***ynthesis - Analysis - Synthesis*** in their money dealing books and talks.

**"YOU ARE WHAT YOU READ AND WHO YOU ASSOCIATE WITH!!"** *(Jim Rohn)*

All the people with exceptional money skills offer not only very peculiar shot-cuts to wealth, but they also instill in us the necessity ***to put the conscious strain on our financial brains.*** The pointI am making here is connected only with the absolute necessity for everyone nowadays ***to acquirefinancial intelligence*** to be able to intelligently manage their lives money-wise. Disrespect for money, innumerable credit cards, and mindless spending result in **SELF-DEMAGNETIZATION**that pushes a person off the ***Self-Actualization*** track and often destroys a person altogether.

*Reminder:* **"Drink the juice of life, but do not drown in it!."** *( Sadhguru)*

There is a British joke about a frivolous son who sent a cable to his dad from college: The joke teaches us how to be both mindful of money-management and how to teach our kids this essentialskill in life. The son sends a cable to his dad: "***No mon, no fun, your son.***" The father responds:"***Too bad, very sad, your dad.***" The message is clear: **"Be a programmer, not the programmedone."** *(Robert Stone)* Please, continue working on your financial intelligence auto-suggestively, using ***the Auto-Suggestive Meditation***.

**To self-subsidize, I need to know who I am money-wise!**

***I am a strong, brave, calm, and determined owner of my firm will!***

***I can** manage my money, **I want to** manage my money, **and I will** manage my money!*

***Money is the Exchange Force, Not My Boss!***

**Become a Much Wiser Money-Actualizer!**

## 3. Good Financial Metabolism = Financial Fitness!

***Financial fitness*** as any other fitness is the result of the metabolism, or ***financial metabolism*** of the mind that needs to be boosted in the form of quality of knowledge to consider and the ability to process knowledge regularly through ***the holistic grid*** in all the five levels. ***Your investmentsinto your physical, emotional, mental, spiritual, and universal*** departments of life on the personalor business level must be reviewed holistically.

Next, ***your connection to the Universal Intelligence at the universal level*** should be working permanently, since you know that you need to tune your **AUTO-ANTENNA**, your intuition on tothe Master Mind and never get disconnected from it.

**The Law of Attraction is constantly in action!**

Each level, in turn, should be processed through the holistic formula: ***Synthesis-Analysis-Synthesis*** The next, the most vital supplement for the financial metabolism is the **SENSE OF MEASURE!**

*Self -Induction:* ***A sense of measure is a great treasure!***

You should not make the desire to get rich, or to impulsively get what you do not need overpoweryour mind. *"Do not be over worn with the desire to be rich, because of your own understanding cease". (Proverb 23.2)* ***People who lack internal fortitude often become victims of the money chase*** or victims of those who have self-discipline and self-management skills.

*Self-Induction:* **Conduct a Self-Limitation war; Less is More!**

Financial fitness needs the ***sense of measure*** to become a part of your financial skills training. One can hardly become money-independent, without having a ***peculiar sense of measure.*** The French feminists say, "If a woman ***has no sense of measure, she has no brains."*** The lack of thesense of measure, as the consequence that any mindless action must be irradiated in us, and the work with a robot-friend , programmed in a watch, or a wallet, will be very helpful in this respect.

We all need to train our aware attention to gain **FINANCIAL FITNESS** first. In fact, this trainingfor financial fitness should start at school. Kids absolutely must get the money-managing skills before they leave high school.

Economics is the subject that is as indispensable in children's education in mathematics. ***Money does not favor those who deal with it casually, without due awareness.*** A financially intelligent person is the one who has learned the rule below:

## Turn Your Intellectual Quality into Quantity!

# 4. Financial Awareness is Your Life's Mast!

It is no surprise that **FINANCIALLY COMPETENT** people are very well-organized and are very personable. There is a certain ***magnetic field*** around these people. ***They glow with the moneyenergy and financial intelligence.*** Money favors them because they respect money in the most rational, initiative-taking, and intelligent way. Also, they are very engaging personality-wise, and,most amazingly, they are truly spiritual!.

I personally admire ***Suze Orman*** who is giving us very instructional and valuable advice that helpspeople not only enrich their financial intelligence, but it also builds up the character of a person who wants to make money de facto, not just de juror. Besides, she is very personable herself! *Her personal magnetism* is amazing and very persuasive! She develops in us a peculiar sense of measure that have outlined above, and that financial fitness requires as an indispensable skill.

***Money has a magnetic power that generates self-confidence and doubles the mind power!***

Following the ideas of a personality development by a great Russian psychologist, Leo Vygotsky,I would like to back up my arguments with his brilliant observation that" ***teaching any subject, weshould not focus on teaching the subject itself, but on "*****teaching a whole person*****."*** No doubt, financial education helps to make the self -image of a person complete.

***Money management skills should be an indispensable part of the holistic education!***

Besides being built on self-discipline, charisma, personal magnetism, and a sense of measure, ***financial intelligence needs patience and the feeling of contentment*** to be developed in the brain.Being overly frugal does not work for consciousness raising, either.

*Auto-Induction:* "***Discontentment is poverty even in riches***." (*Confucius)*

In our time, the plague of petty, ***impersonal bureaucracy*** is familiar to everyone. It kills our enthusiasm, down-grades our abilities, and depletes us of the best intentions.

**It's worth every treasure on earth to be strong for oneself!**

But being strong is not enough. You need to be orderly and disciplined, correct in our speaking and controlled in our reactions. Most importantly, you must be financially free ***because "money is not the root of all evil; lack of money is!"***

***In sum, t***o frame your life financially means to have **a STRATEGIC PLAN OF ACTION** and program yourself for orderly, not messy, sporadic thinking, speaking, and money-dealing.

## Money is in a Friendly Seat only if you Intelligently Manage it!

# 5. Money-Actualizing Skills

***The Money Managing Skills should be developed in synch with the Money Actualizing Skills,*** not by way of grabbing some valuable tips over the Internet or trying to just pass the bankruptcy test. Your own philosophy of money-making might be shallow knowledge-wise and experience- wise. ***Financial awareness is also a cause-effect awareness of the money-flow***.

In the present-day ***Informational Biology,*** we can program life by changing the sequence of DNA in a genome code for a predictable outcome. Likewise, ***we can willfully change our financial code in the brain*** and teach ourselves to manage the money to our advantage. Many people have trouble living within their means. Now researchers have found that ***there is a gene, linked to credit card debts***. *(Emmanuel De Nave)*

**"A particular gene affects our financial behavior."**

Apparently, we need more ***neuro-scientific education in this respect***, too, to be able to have morefinancial awareness backed up by brain work awareness. Like synthetic engineering, ***we can engineer money intelligence in ourselves***, changing the DNA of our limited thinking into the unlimited one. Piecing together of the needed mental genomes by using complex intellectual algorithms will help us re-program our cells for better money managing in time and space.

The observation of the main cosmic law will add a special quality volume to our overall intelligence and, consequently, consciousness, and it will improve our behavior in accordance with ***the Law of the Right Human Behavior*** in which money is a very important vector pointing to perpetual happiness only if you make the money work for the good of many or at least some people, not just for the gratification of the material needs of one person.

I recommend reading the book "***Why We Want You to Be Rich***" *by Donald Trump and Robert Kiyosaki* in which they express very vividly why the rich get richer and the poor poorer. The bookproves that knowledge put to action is double power!

In sum, ***financial intelligence will protect us from the inflation of our thinking in the financial area*** when the use of the old patterns of money-monitoring results in financial failures that aggravate our life and hamper the way for self-actualization. ***True, the money God is presiding the world! Empower it with your Intelligence God!*** I am sure that the birth of the biological robotswill help us manage our money in a much more conscious way.

Thus, the technological advancement is gradually reprogramming our cells in an irreversible direction, and our **FINANCIAL MATURITY will** grow with our self-consciousness in unity.

# Let's help the Intelligence God Rule the Future World!

# 6. Dematerialization is the Spirit of Giving!

The spirit of giving is the next important aspect in line. "*Whether born of doing anything with gratitude, revealing in a breathtaking sunset, creating a work of art, making love or just plain being thoroughly decent to a fellow human being, the portion of positivity we receive is no longersubject to entropy and* ***becomes money in the bank***" (*Rev. P. s. Berg*) "***Generosity, like love andgratitude are the sentiments that free and raise individual consciousness.***" (*Piero Ferrucci*)

The one who's giving is also receiving! Generosity integrates the whole psyche around a single focus - ***generating an incredible soul integrity*** that makes the owner of it the happiest and the noblest man on earth A giving person is more likely to develop himself in the spiritual area than the one that does not have such ability. Spiritual people are amazingly generous.

*Self-Induction:* ***To change your bad mood, be in a hurry to do something good!***

Giving is the ability to internalize emotions and externalize the mind. Carl Yung when asked whathuman quality he deemed to be the most important in a man replied that it was a person's attitudeto money. He said, ***"If a person is cheap in dealing with money, he will be cheap in love or in anything he does because the spirit of giving the best of him is not there."***

Let's remember with a heart-felt warmth of thankfulness and pride that the USA is the greatest giving country in the world! There is the greatest number of people here who are generous enoughto donate money, their valuable possessions, or pieces of art to our universities, museums, libraries,theatres, schools, etc. Millions of dollars are sent to the countries in need all over the world! **The spirit of giving is in the country's logo!** However, mindful, compassionate, and considerate giving should not be confused with the giving to buy someone's service or help in return. A lot ofgiving is motivated by the economical o, political, or egotistic reasons.

**Generous gestures of people make their conscience more illuminated!**

Giving is, in fact, the best exercise to establish the **MIND + HEART** connection. It is the primarydemand for your self-growth and raising your self-consciousness and cleanse your conscience.

**Any giving is Conscience and Love-Refilling!**

With unconditional giving, your life becomes richer, more beautiful, and harmonious. Giving loveis our best privilege before the humanoids. Their hearts won't be filled with an attitude of gratitudefor someone's sincere gesture of a gift Any generous gesture is rewarded with a happy smile thatraises our self-consciousness and self-respect immeasurably.

## In My Mind's Store, Getting means Giving More!

# 7. Financial Consciousness Inflates Your Pride with thePersonal Integrity Might!

***Financial consciousness*** is a vital part of our overall consciousness! Every sacred book teaches us to give, to share, to contribute. ***We all know that, but we are not aware of that***. The insecurityof life, and the uncertainty on the job market generated ***more impersonal behaviors*** on the part ofpeople in our social service offices .in which opening the mind and the heart to a person's concernis hardly the case. To know a person better, *Carl Yung* suggests we pay attention to his / her attitudeto money. He writes, "***If a person is cheap in money-dealing, he is cheap in love, in friendship,in a family, in any undertaking"*** because there is no soul width in his character midst. Many people seeking help feel belittled, and such situations generates stagnation in self-consciousness growth. The necessity to admit it, or the need to ask for it humiliate a noble soul.

**Nothing makes us smaller than the lack of money!**

We need to change the pattern of our victimized thinking and compulsive shopping habits that result from financial insecurit*y* when a credit card becomes your best helping hand **and** install new, constructive, self-reliant, resilient, responsible, electronically-backed up and strongly-willedpatterns of money management, focused on our renewed financial intelligence, monitored level by level**.-** ***physical, emotional, mental, spiritual and, universal***. We should follow this impeccableorder of things in life in terms of money and its impact on our personal growth because it is moneythat often impedes it. We all marvel at the way people manage to make a lot of money, but we rarely learn from their ***exceptionalfinancial intelligence*** that expands their personal universe. Asis indicated in *Thesis 2, in the Introduction,* before you can get something, you have to become somebody. The money-making ladder was, is, and always will be:

**Be someone!** ⟹ **Do something!** ⟹ **Have what you want!**

Also, the person, ***committed to Self-Installation in life*** wouldn't let anyone script his actions for him, or monitor his money, without his full awareness of how and why this is being done. ***The financial consciousness is integral and complete.*** **Its conscience is in the form +content link.**

***Form*** + ***Content***

**(Body+ Spirit+ Mind) + (Self-Consciousness + Universal Consciousness) = AnIntellectually Spiritualized and Financially Aware You!**

# Financial Consciousness means Managing the Money, Not Letting the Money Manage You!

# The Intellectual Odyssey of a Seeking Soul

*Vistas of Intelligence to Install Next:*

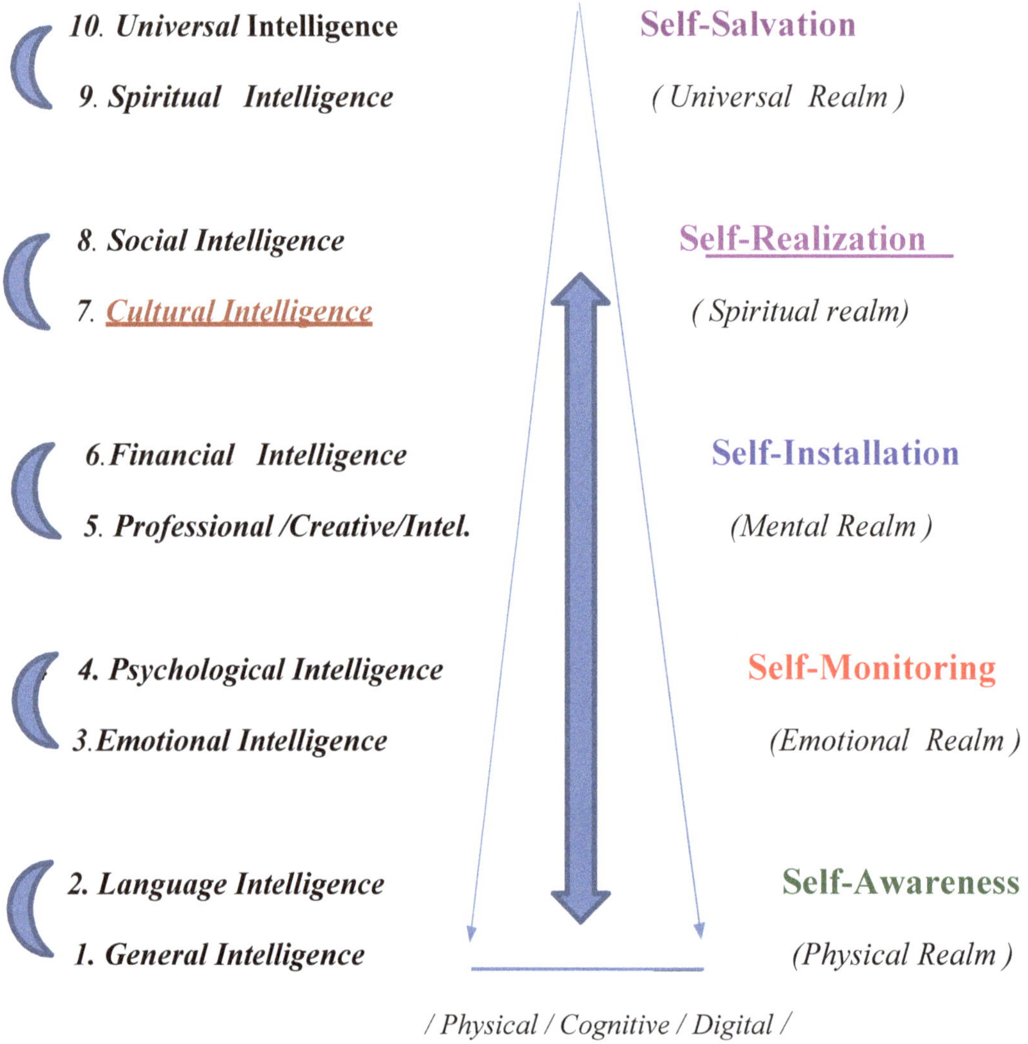


**We Are All Structured in the Universal Culture!**

***The life on Earth is beatified with different cultures' might!***

## To Make Your Life Culturally Wise,

## Generalize -Analyze - Internalize - Strategize – Actualize!

# Stage Seven

*Macro Level - **Spiritual Dimension** – **Self -Realization***

# Cultural

# Intelligence

*Change the Cultural Gravity of Your National Vanity!*

**In the Global Culture,**

**We are Universally Sculptured!**

# Cultural Intelligence without Negligence!

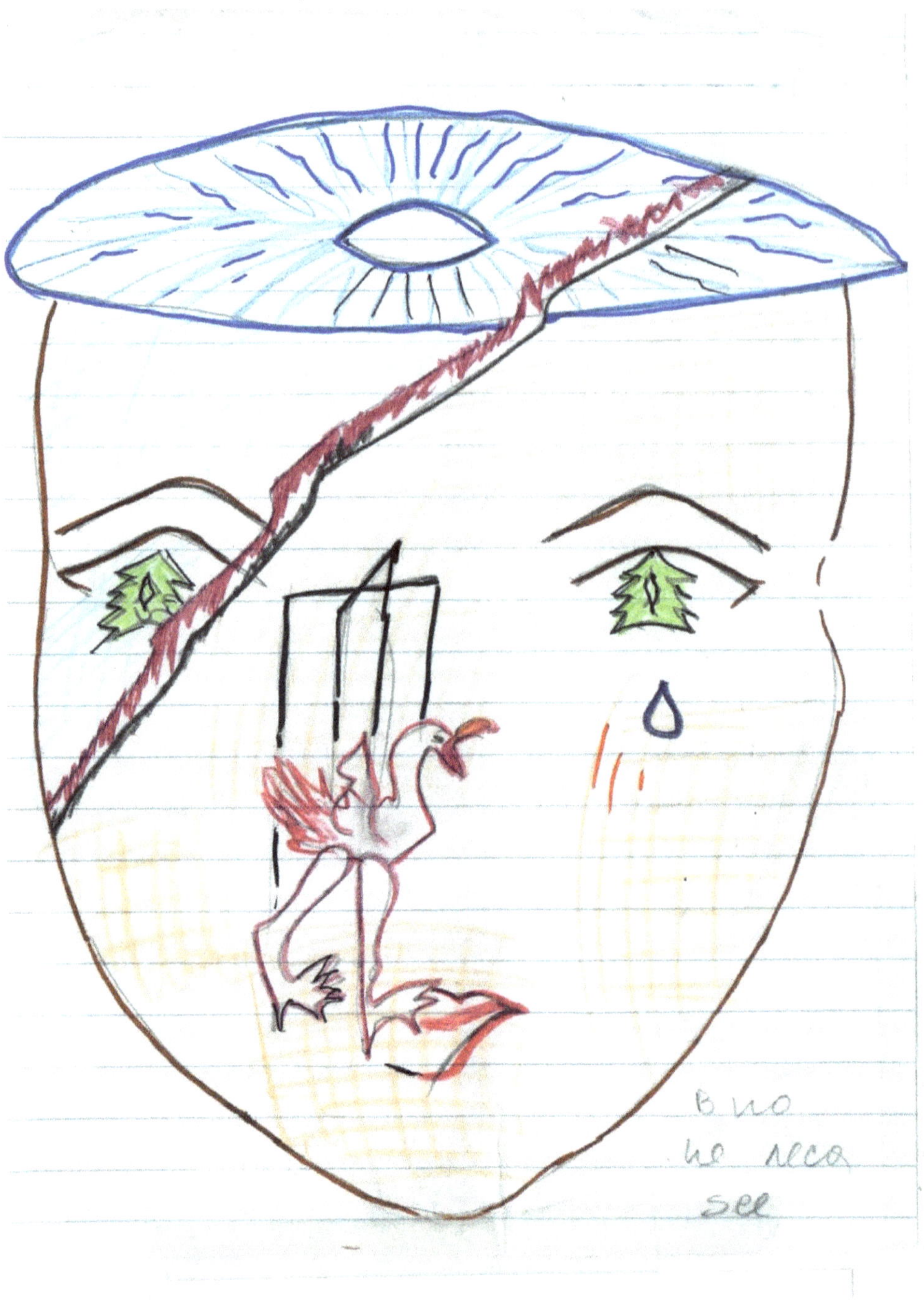


# In the Cultural Quest, I Manifest My Best!

# 1. Cultural Awareness as the Code of Behavior!

Having developed, or at least enriched your overall intelligence in the three main levels: ***Mini*** *( General and Language Intelligences),* **Meta** *(Emotional and Psychological Intelligences),* **Mezzo***(Professional/ / Creative, and Financial Intelligences)***,** we need to explore the ***Macro-level – Cultural and Social Intelligences.*** These are two levels that ***define us as personalities*** with real integrity because this is the spiritual level**.**

***Cultural intelligence is the ability to adapt to a new cultural reality.***

**The Absence of the Cultural Intelligence characterizes our political turbulence.**

***Cultural Intelligence*** is what constitutes culture and how cultural variety and vibration lead to an understanding of how these variations affect our behavior. Different cultures have different values and different patterns of behavior and to be an intelligent person that wants to climb the social ladder, you must embody those values, or at least to be familiar with them in any socialenvironment.

**We Are One of a Kind in the Universal Mind!**

Recent trends, including high-tech communication, have created new challenges for globalmanagers**.** They involve three components: ***knowledge, mindfulness, and cultural awareness.*** Cultural awareness needs to be a part of the whole as **our new moral code**, one for all nations, the **INTERNATIONAL CULTURAL CODE.** It provides the essential awareness in our everyday interaction on the global social arena that is the Internet now and the AI invasion.

The international code of behavior is developing a common cultural structure, and it will unite allthe nations to the point that cultural respect will be viewed as the main code of behavior for the humankind hopefully in the nearest future. The spirit of **CULTURAL INTEGRITY** must be instilled in humanoid, too. ***Cultural awareness leads to cultural acceptance and heightens a person's consciousness*** that should b*e* **SELF-MONITORED** and possibly AI supported now.

Cultural and social intelligences are in the spiritual dimension because our cultural and social perception needs to be expanded to the spiritual level when we respect just the mere fact that we are all human. Such approach demands attention to the subtle cues from colleagues and clients who may possess different cultural values. ***We need to let go of the cultural and color-blind prejudices*** to simply understand how culture influences attitudes and behaviors with full awareness. **Our robotic industry must be international in its essence.** There are many culturalcodes in the world, and we must be familiar with those that we get in touch with, respecting themfor their uniqueness, engraved in a personal individuality.

# Globalization of the World Connects Us by One Digital Code!

# 2. In Digital Life Perspective, We Must Be Culturally Introspective!

Life demands we guide ourselves by new ***ethical and moral codes.*** We mix mostly on the physicallevel, but hardly ever on the emotional, mental, and most importantly spiritual levels. There are still many ignorant, culturally and religiously separated people around the globe who scorn ***Cultural Intelligence*** and call some international students that are into deep learning "***egg heads or nerds,***" while such students constitute the best part of our education and often self-educate themselves to the level much higher than any local school can provide. ***Bulling stems from such close-mindedness due to the lack of cultural education in our schools.***

**Cultural Intelligence must be upgraded at every level of our Self-Creation!**

Terrorism in all its ugly forms is also the result of the lack of cultural intelligence. There is a lot of cultural diversity around us, and very many people are incredibly talented. Does God bless us with the gifts of exceptional mental value because we are *Jewish, Chinese, Japanese, British, Russian, or Iranian?* We are loved and blessed equally, and we need to finally recognize that andvalue intelligence, not a national identity, as the highest commodity ever Cultural education expands the horizons of the beauty of life in the kaleidoscope of its varieties. ***It is an essential ingredient of a person's conscience and his growing self-consciousness!***

We live in the global culture, now, and globalization demands we establish, **mind-to-mind and heart-to-heart connection** among cultures. The diversity of the world's cultures enriches our **COMMON INTELLIGENCE** not only in schools, but in any administrative institutions-, policeoffices, and different courses where people must analyze intelligently any problematic situations for their cultural ingredients. ***Being culturally intelligent helps resolve the conflicting problem inevery area of life***, starting with the international affairs because lack of cultural awareness traumatizes people and leads to ***the stereotyped solutions of the cases*** in courts, or even on the higher governmental levels. **The value of a human life must be restored** to welcome transhumanism that is our common evolutionary fate.

The key to becoming culturally intelligent is ***the awareness of the conflicting values andperceptions across cultures***. Any culture is a rich source of language, art, music, fashion, food, ideas, and incredible talents of people that contribute to the world's beauty. Cultural education isessential in our schools because it prepares students for business communication on a global scale.

It teaches them to better understand the needs of the customers from different parts of the world the knowledge of which enriches our general intelligence beyond belief ***and evolves our commonconsciousness to a new, higher level of self-consciousness*** that becomes culturally enriched.

## We are of the Same DNA – You, Me, and They!

# 3. Cultural Metabolism and National Inflation

***Cultural Intelligence is the ability of a person to adapt to any cultural reality and respect it for its uniqueness***. It is, in fact, cultural ***metabolism*** that allows culturally diverse relationships coexist. Unfortunately, we often deal with **cultural inflation** or **cultural ignorance** when the control drama prevails, and cultural differences hinder our evolutionary progress. **CULTURAL STAGNATION** slows down the metabolic processes in our common social body that is digitallyenergized now. We do not discriminate humanoids for their cultural and racial peculiarities.

Humans, though, remain technologically advanced, but culturally ignorant! The gap between different cultures and the ignorance that people display even about the geographic position of manycountries on the map is appalling! When 28 years ago, I came to t the USA from Latvia, one of my colleagues thought I was from Africa, confusing Latvia with Libya.

**Acceptance of other cultures comes only with the Cultural Intelligence!**

People love travelling and tourism, and they undoubtedly contribute a lot to our cultural integration. But cultural intelligence is not limited by the outer impressions, it is the change of programming in the brain and ***the establishment of the synergistic connections*** in our cells. The ***cultural metabolism*** should be also working on the auto-suggestive level of mutual acceptance

**We are of One Blood in the Universal Gut!**

Each culture, however small, has enriched the world culture, and our kids should know what contribution their culture has made to the global cultural beauty and be proud of it. ***Knowledge ofwhat constitutes culture*** and how cultures vary leads us to an understanding of how these variations affect our common intelligence that is forming our **common vision of the world** as ourcommon home. Our behavior patterns, business growth, politics, and the destinies of entire nationsgetting integrated now. The rule, ***"Do it my way, or no way!"*** does not work anymore.

In sum, **CULTURAL AWARENESS** generates healthy cultural metabolism that is the acceptanceof another human being's differences to every one's satisfaction, and it is a sure way to resolve any conflict. Unfortunately, we continue controlling each other ***without an objective observation of the cultural peculiarities*** of our colleagues, clients, or even relatives. ***Cultural inflation generates religious bigotry, hate, and violence.*** A person, lacking in cultural intelligence tries tofit everyone into a habitual framework, judging people by his own limited standards. Fortunately,our mesmerizing AI creations in different countries and in various nationally humanized forms demonstrate our **HUMAN ONENESS**, teaching us to appreciate it and marvel at it.

## Culturally Limited Perception is the World Must be Digitally Re-shapedand Re-thought!

## 4. Compatible Brain Chemistry

The development of cultural intelligence also demands that we create a **COMPATIBLE BRAIN CHEMISTRY** in our social relationships. It means that ***aware attention*** should be paid to everyperson that you meet in life because ***every person is on the same life path with us for a reason***. We need to unwind the stereotyped perception of people and unblock these spots in the mind! **Inthe digital world, no one is superior to anyone!** Our interaction on the computer or through thesmartphone befriends us much quicker that in our face-to-face communication.

We differ only in the level of our awareness in electronics that widens our outlook and general knowledge of the world thanks to our connecting mass media giants- ***Google, Facebook, LinkedIn, YouTube*** and others. Once, you master these global media outlets, you open your **mind + heart** connection to the world. Socialization of our lives on an unprecedented scale and the opportunity to communicate with people from different cultures gives us a lot of food for thought. We have a rare chance to train ***the right brain hemisphere***; ***developing an objective communication*** with people around us, as opposed to *the* ***subjective communication*** that is self- centered, egotistic, and limited. **Our brain compatibility is amazingly uniting now.** Hopefully, our future will be based on the God-given basis of human objectivity.

To solidify the process of **INNER UNIFICATION**, ***"we need to identify partners with compatible brain chemistry."****(Helen Fisher, Discover, March11, 2013)* In the new reality, to sizepeople up intellectually, financially, professionally, emotionally, ***we need to be more culturally-informed tolerant*** from the objective premises. It does not mean that we should be more discriminative, it means we need **to adjust to the digital era in five dimensions**, remaining ***noblyhuman, compassionate, and respectful*** to those people that will be swept off by the digitally enhanced industrial upheaval due to the lack of education and digital skills.

*Auto-Induction:* ***Don't size up, open up, and help up!***

Let's practice patience of mind and tolerance of feelings. Let's filter our thoughts, words, feelingsand actions at work and at home, in public, and in private for their common values and get unitedon their basis. Let's defy the gravity of cultural ignorance and religious prejudices! Don't polluteyour mind and heart with the thoughts and feelings of scorn, superiority, or indifference. Every person, no matter how different culturally he seems to be, is a human being that is ***yourcontemporary in time and space***. There is a proverb *in Russian* that tells people to be in a hurry to do something good to a person who is culturally unfamiliar to you. ***"Be in a hurry to do something good." Kindness is our friendship hook!"***

# Assess a Person's Fairness by his Cultural Awareness!

## 5. Establish a Synergistic Cultural Climate. Be Self-Refined!

We sometimes experience some hostility at work on the part of someone that is ***culturally unaware.*** We might manage to get out of the back-lash loop successfully, but if an unpleasant situation is generated by the leader of the group, a boss, or a supervisor, the presence in such an environment becomes very stressful and even harmful to our health, to say nothing about the quality of work that we might be responsible for. Doing any work becomes a real burden, and wewish we could avoid being in the same room with ***the person of emotional abuse*** that we dependon, anyhow. Emotional intelligence needs to be applied here to change the vibrational quality of the energy in the room.

**Generating a pleasant, synergistic climate at work is the rule # 1 for any manager.**

Being a right-brained, creative, positive, content person goes beyond personal good. ***It works for the common good!*** Such attitude helps generate the feeling of welcoming peacefulness, respect, and trust that we are all seeking in life and without which our lives are not complete. So, let's establish ***the synergistic climate and a sense of Oneness*** with every colleague at work or every member of the family at home because we are dependent on one another. We see a lot of such cultural integrity in different sport teams that demonstrate to us that we are the citizens of the planet Earth, and our national belonging to this or that country is secondary. ***No doubt, interdependence is the basis for any job, any business, and any love relationship.***

This interdependence is not only professional; *it's intellectual, emotional and, psychological*. It'sgood ***to be an independent thinker***, an individual, but not ***an impersonal individualist*** who caresonly for his/ her own persona, for the paycheck and the security of retaining the job at the expenseof being compassionate, understanding, inspiring, responsive, and responsible for another humanbeing's ***professional comfort*** at work. Unfortunately, it's appallingly common these days.

**An impersonal attitude ruins relationships, and eventually, it damages our lives.**

Alongside with the cultural awareness, ***we need a lot of sensitivity, empathy, and tact*** to share thespace and time with the people that life might place us. In. It's remarkable how we are all part of**ONE LIFE SYSTEM** - ***one life system!*** So, interdependence is what keeps us together and makesus strong to evolve. Synergizing your inner space by means of the auto-suggestive work, you willmanage to synergize the outer space, too. ***The people around us are our contemporaries***; we share the same space and time with them. Our life is terminal, and this axiom must make us appreciate life in ourselves and the people around us. A great Russian poet, *Boris Pasternak* wrote,

**"The Miracle of Life is just an Hour Long.**

**Live this Hour Honorably before You are Gone!"**

# 6. Mind + Heart Link Must Always Be in Sync!

It's essential to control oneself first and establish a synergistic, productive supervision of the people under you, with a welcoming smile, ***a supporting word of praise***, and an honest appreciation of someone's contribution to a common course at work. Unfortunately, our sincere and sacrificing work is often taken for granted, backed up by impersonal comment, ***"It's your Job!"*** So, learn to compliment the people around you honestly and wholeheartedly.

*"I can last on a good compliment for a week." (Mark Twain)*

Be sure to synergize your relationship with timely and sincere complimenting your colleagues andthe loved ones to make our mutual life sunnier and more pleasant. Being synergistic means beingable to generate good, friendly **heart + mind** fortified vibrations around us, helping people that share your time and space with you feel calm, inspired, and appreciated. As *Albert Einstein* said,

***"Happiness is being content with the space you are in and the time, allotted to you."***

If you have an argument, learn to channel the discussion by ***the method of consensus building*** that helps people working at the same project gravitate together. Consensus building is synergizing the working atmosphere and helps people transcend the uncomfortable feelings of fear and the expectation of a conflict if he / she voices out a different opinion of any problem.

**The spirit of consensus is the spirit of liberation and joint reformation.**

We begin feeling free from ***the pressure of competition as a looming force of elimination***, and we start seeing competition as the means for constructive self-growth and illumination. Inclusiveness is needed in the training of humanoids, too. They should be programmed for the respect of our humanness, not be aggressively disposed towards it.

**"I am One and you are One; we are One together."** *(Tara Rae)*

If a person is argumentative, if he / she is challenging your ***cultural identity*** and even uses the derogative language, immediately surround that person with brilliant white light in your mind. I am sure that the light will enlighten that person with the realization that he / she might be wrong. ***Surprisingly, it works!*** If you are the one who generated the conflict, it's your responsibility to acknowledge your misgivings with the light thought in the mind. ***When we get racial and nationalistic, we become disconnected and sadistic!***

*Auto- Suggestion:* **Be kind to the unkind, Be One of a kind!**

# Establish a Mind-to-Mind and Heart-to-Heart Communicationwithout Frustration!

# 7. Be Culturally Apt Auto-Suggestively!

The ability to influence and persuade others can be developed only through love and patience, through respect and acceptance of another person. ***The ability to tune to the feelings of the peoplefrom other cultures, to feel empathy and enthusiasm*** as opposed to the feelings of superiority, rigidity, and moodiness is easier obtained through autosuggestion.

**I know who I am, and I know who I am Not!**

**I let others to have the sense of self-worth.**

**I can** *accept some one's difference* **I want to** *respect some one's values.* **I will *become*** *aculturally intelligent person.*

Now that English has become the universal language, cultural intelligence should be an essential part of a person's education. Our courts will be much fairer and more considerate in their judging,and people will be more harmonized in their inner responsive framework and, therefore, they'll bemuch more productive at work. ***Lack of cultural aptness is a real sore in a person's individual score!*** As a matter of fact, our intellectually spiritualized fractal of self-growth is the same.

**The Fractals of Intellectually Spiritualized Beings:**

***Form*** + ***Content***

**(Body+ Spirit+ Mind) + (Self-Consciousness + Universal Consciousness)**

= A new Intellectually Spiritualized and Culturally Aware You!

At this level of developing ***intellectualized spirituality***, cultural intelligence entails the necessity to instill in our kids and college students the **CULTURAL BEHAVIORAL SKILLS** so our education could further ensure smooth and effective ***cross-cultural relationships***. We should givethe United States a great credit for uniting people of different cultures and changing their mentalityfrom a limited cultural one to the international and eventually, **GLOBAL OUTLOOK.** People stop being *Chinese, Italian, Russian, Hispanic, Turkish*, etc. here. They become just the citizens of the planet Earth, *like the Humanoid Sophia*, ***and that is the mentality we are all heading toward.***Isn't it very promising? So, let's instill in our minds the mind-set that says.

*Auto-Induction* ***Be Culturally Apt; Culturally Respond, Don't React!***

# Every Human Contact is a Responsibility!

*(End of Stage Seven- Cultural Intelligence / Spiritual Dimension)*

# The Intellectual Odyssey of a Seeking Soul

***Vistas of Intelligence to Install Next:***

**We live at the age of an unprecidented socialization!**

| | |
|---|---|
| *10. **Universal*** **Intelligence** | **Self-Salvation** |
| *9. **Spiritual Intelligence*** | *( Universal Realm )* |
| *8. **Social Intelligence*** | **Self-Realization** |
| *7. **Cultural Intelligence*** | ***( Spiritual realm)*** |
| *6.**Financial Intelligence*** | **Self-Installation** |
| *5. **Professional / Creative/Intel.*** | *(Mental Realm)* |
| ***4.Psychological Intelligence*** | **Self-Monitoring** |
| *3.**Emotional Intelligence*** | *(Emotional Realm )* |
| ***2. Language Intelligence*** | **Self-Awareness** |
| ***1.General Intelligence*** | *(Physical Realm )* |

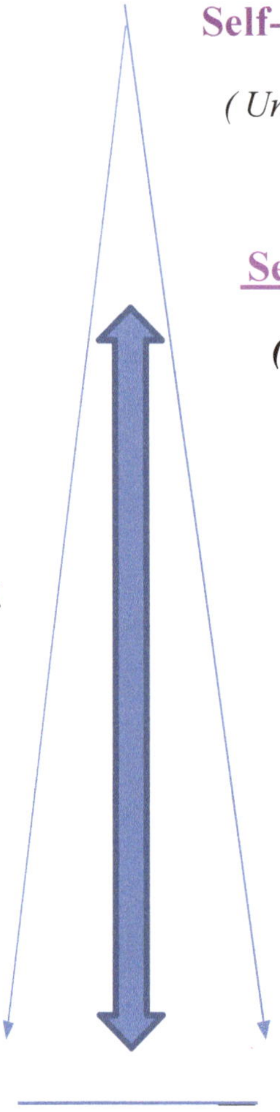

*/ Physical / Cognitive / Digital /*

**Stick to the paradigm-Synthesis-Analysis -Synthesis line!**

**Inflate Your Personal Pride with the Social Might!**

## To Make Your Life Culturally Wise,

## Generalize -Analyze - Internalize - Strategize – Actualize!

# Stage Eight

*Macro-Level* - ***Spiritual Dimension*** - ***Self -Realization***

# Social

# Intelligence

*(Upload your smart phone with a new Interpersonal Intelligence Tone)*

**In the Universal Array, we are All just Human Clay!**

# Our Social Big Bang is in the Unity of Ying and Yang!

**Resist the National Vanity in the Social Gravity!**

# 1. Social Awareness and Human Aptness

***Social Intelligence*** is the path of our obtaining global knowledge in cosmology, genetics, and in any branch of science that unites us socially. It is an integral process of a gradual self-transformation in five dimensions, a considerable enrichment of spirituality and s **SOCIAL AWARENESS.** ***"Humans are inherently social beings; that's our evolutionary legacy.*** " The quality of our lives now is the quality of our Social Intelligence or our global, digitally enhanced vision of the entire world.

## Anti-social behavior is the lack of intellectualized spirituality.

Evolution through relationship shapes us personally and socially. No doubt, socialization promotes our intellectual culture. However, never have relationships been so problematic and conflict- bound as they are now. "***Our collective reality is a symbolic expression of fear and an insatiabledesire for security, as well as heavy layers of negativity and judgmental attitude that we have accumulated in our collective human psyche.***" *("Scientific American Mind," (July 2014)*

## "A shift in planetary consciousness is taking place now."

As any other intelligence, ***social intelligence also needs to be holistically developed*** because it incorporates all the vistas of intelligence outlined above. An unenlightened person usually has a narrow vision of himself and life as if he were looking at the world through a small window. However, when a person expands his self-consciousness, it is like going through a door into a world of communication. Every one of us is globalizing his mind, the vision, his / her personality,and a professional outlook. "***Instead of fighting the darkness, bring in light. Allow all people tobe as they are. Shine for others, too!***" *(Dale Carnegie)* Being ***socially apt*** means establishing **a mind-to-mind** and **heart-to-heart** connection with all the people you get in touch through a face-to-face digital communication. If we remain impersonal, we cannot have a peaceful, healthy society, ***based on intellectualized spirituality and devoid of any religious restrictions***. Learning to expandour Living Intelligence holistically, we are also developing new respect for life in yourself and others, including thee humanoids because our neurological circuits are implanted into them.

The vision of ***a person or a humanoid who has obtained the living intelligence in five levels,*** becomes inwardly stronger. The mind performs correctly and in unison with his trained emotions.People gravitate to him / her for spiritual energy and life wisdom. At this spiritual level of self- training, you are also obtaining the **Ultimate Result Vision** of yourself. (*URV*). You are changingthe ***form and content*** of your life; you are becoming ***the Vintage you.*** Having read the book that far, ***I*** can vouch that you have developed a solid trust in yourself, and you can handle any situationconsciously. You are at the point of self-revelation and self-realization! You can declare<

# "I Am Free to Be the Best of Me!"

# 2. Social Metabolism without Any" Ism"

Forming the ***holistic content of your life and*** propelling yourself forward with ***a much more refined self-concept, you*** become socially apt, more responsive, and responsible, and that is a greataccomplishment of our unprecedented socialization. Changing ourselves, we create a new ***holistic WE*-CONCEPT** of social unification and mutual responsibility for the life on the planet Earth.Our new digitally generated social connection is gradually forming a new world community of **GLOBAL CONSCIOUSNESS**. ***Information is ammunition because our awareness of it changes the content of our lives!***

Unfortunately, due to high tech explosion, life has a different narrative of self-erosion! W***e live ina socially*** and digitally polluted world. the world of social labels with ***the "ism" endings*** of all kinds. It has become an urgent necessity to disintegrate ourselves from the social pollution and formulate our own personal standpoints, finally ending our dependence on *"**the collective unconscious,** "**religious limitedness, and nationalism.***

**"The cheapest pride is the national pride**!" *(A. Schopenhauer)*

*Our* **SOCIAL METABOLISM** is now clogged more than ever with impersonal ***egoism, nationalism, chauvinism, and racism***. Many people do not think on their own; their judgement is formed by the **crowd mentality and the mass media indoctrination.** Mass media does not provide a holistic, ***objective view*** of the circumstances that generate the events.

**The Law of Cause and Effect is not observed by our mass media!**

Being in the dark, people often adapt the commonly programmed view. Total unawareness, woundup by the manipulators of the minds, generates ***impulsive reactions*** of the *"collective unconsciou*s," perceived as the fight for justice that, in fact, is the struggle against a holistic, insightful vision of life. ***"Our seeing of the reality requires a correction of mind; just as clear vision requires a correction of the eyes*** *" (Alan W. Watts)*

The content of each individual life needs to be sifted off its polluting stuff! Negative people, likeredundant information need to be sorted out by you. Our interdependence on the tools, manipulating the social opinion, entangles the. minds and blurs ***the objective social vision***. So, let's be governed by aware attention and ***purposeful sorting out of the social contacts*** that we have. We have no luxury of wasting time on the people of no intelligence and no human value. Our social circles will be expanded by thinking machines and socializing with them will become a great treat and an incredible learning experience.

## Socializing means Getting Interested in Other Cultures and Becoming More Interesting Yourself!

## 3. Mind-to-Mind + a Heart-to-Heart is a New Social Art!

***Social Intelligence*** is at the top of ***the Self-Acculturation pyramid.*** To learn social skills, a personneeds to incorporate all the above studied intelligences and become a new human being - a reallyadvanced one, a man / a woman of integrity, ***able to combine thought and feeling*** and be sociallyfit spiritually. A higher level of interpersonal mental frequencies is needed. ***A new level of social vibrations is changing the world.*** And now, we need to come to terms with the human-robots thatwill help us get rid of our religious boundaries because they have none

**Our Mind-to-Mind and Heart-to-Heart is becoming a digitally enhanced Art!**

Any art is a form of communication, but this form becomes classic if a piece of art enclosed in it radiates beauty, has a structured composition, and is very thought-provoking. I think that our socializing in general and the one on the Internet should be much better both ***in form and in content*** and our Internet companies should contribute more to the quality of our socialization.

***The vital principle of holism is ruling the world of unification in human salvation!***

Amazingly, socialism was uniting people, while capitalism is dividing us. Obviously, the best features of both social forms are to be adopted by us on the new twist of the evolutionary spiral. The countries that have incorporated ***the features of both social systems*** into their national framework are doing much better than the rest of the disconnected world. *Sweden and China* are the best examples of such conversion that results in a much better standards of living.

***The concern for the well-being of the people will be driving us toward unification,*** and that's theonly way for all the humanity to survive. Social behavior has evolved beyond belief thanks to social media. With the younger generation leading the way, large groups empowered with Internet- enhanced communication accomplish a variety of diverse goals. **We are gradually becoming global software entrepreneurs**, and ***we develop new electronic competences*** of holistic value every day. The beneficial effect of ***Singularit***y adds to our social evolution, cleansing us of any centuries accumulated religiously divisive pollution.

**We enhance our souls and expand social outreach with a digital switch!**

However, ***our social communication should never lose its individual characteristics.*** Often, ineptsocially people do not integrate collectively and are believed to be anti-social. However, these people are better thinkers, and they can contribute more to the society than the ones that merge completely with the massive culture and lose their individuality. In fact, we get back here to ***the first level of intelligence – the physical level.*** It's impossible to help and love others, without firstdoing your own self-quest and giving your best to the society.

## The Identity of the Hearts and Minds Binds!

# 4. Social Network Equilibrium

The difficulty some people have in global communication comes from their ***different mental wavelengths*** which are not coinciding with the vibratory frequencies of the people they communicate with. *"**It is as if we tune in to a certain broadcasting station on a wrong frequency**."* *(John Baines)* A mind-to-mind communication becomes possible only between the people of adequate intellectual frequencies that decisively influence ***the quality of*** **CONSCIOUS COMMUNICATION**between people. **Holistic Intelligence** removes vibratory disharmony and establishes a social equilibrium of a mind-to-mind contact of mutual understanding without being politically bereft with a war-mongering mentality.

New social skills that we need to develop at this level should not conflict with a sincere intentionto contribute to the world our unique gifts of intelligence. With the unprecedented speed of life that stresses us out beyond belief, we need to establish a mind-to mind and heart-to-heartconnection between people that we communicate with for business's sake or personal reasons.

Our new social experience proves that *aware attention displayed on the mental level and the emotional levels*, backed up by the cultural intelligence ensure much success in business solutionsand ensure that a person's individuality will not get lost in the ocean of technological havoc. The holistic paradigm (***Synthesis- Analysis -Synthesis)*** needs to be followed religiously in fivedimensions for the solution of any social problem.

## Generalizing-Analyzing-Internalizing-Strategizing-Actualizing!

The contribution of our computer mega companies, like *Amazon, Face Book and YouTube, LinkedIn* is huge, and the improvements they are implementing on this path are most promising.The social pollution puts us in front of the necessity ***to X-Ray the social contacts*** for their personal,emotional, mental, and spiritual validity for us. **People that we allow into our lives shape our lives in every dimension**! The dynamics of socialization at present are extremely and ruthlessly competitive, but our social metabolism should grow intellectually and in a holistic way. "***What makes the meaning of life is people. So, try to be good to people immediately around you and ina broader, world community.***" *(Reid Hoffman and Ben Lashocha)*

However, if you observe unconscious behavior in your partner, ***hold it in the loving embrace of your knowing, so you won't react***. You need to respond consciously and intelligently and cut theperson off your social loop then and there. ***Intelligent networking is extremely beneficial for self-creation.***; Learn to establish it. *The Scientific American Mind", Nov. / Dec. 2018 writes,*

**"A single brain is a computing marvel, but it pales in comparison to the powers of the net-worked mind."**

# 5. Social Connection is Language Interwoven!

For our digitized social communication to be more harmonized, we need to become ***more polishedin our language competences***. The correct use of the language, the tone of the language used, andthe vocabulary chosen to get the needed thoughts across are crucial now. *(Check out LanguageIntelligence above)* It's good that the use of profanity is being checked by *Facebook*, but we needsuch attention to be paid to the quality of the language we operate with on every sight. A human being at the other end of the virtual line is as real and perceptive as you are. Both verbal and digitalcommunication demand your ***thinking first then saying something or writing a response.*** We should not generate negative consequences for ourselves or other people due to our texting on impulse and transmitting this impulse to millions of others. Again,

**Every human contact is a responsibility!**

Remember," *90% of people do not care about your problems, and the other 10% are glad you have them*" *(Carl Yung)* A great psychologist means that we need to be centered on confronting our own problems ourselves," being ***the thing in itself*"** in solving them because being part of all life, you are still solely responsible for your own life and its outcome. ***"You cannot conquer life because the part cannot conquer the whole, but you can be the best part of the whole!"*** *(Carl Yung)* "Disconnection is death!

**Our new Self-Consciousness formation is the Salvation!** Being a part of the whole is also speaking the language of the whole, called "***Globish***." English needs to be respected for its ***unifying quality*** and the incredible ***feeling of belonging and being One in the context of life***." The information being programmed in English is the greatest accomplishment of the Internet!'

***The dawn of human brain-to-brain connection has arrived with English"*** *(Rajesh P.N. Rao)*

So, let us not distort this unique language while texting, speaking. or writing. To grasp the complex problems of socialization with its political conflicts and war-inflicted sorrows, we must bring about the understanding of ourselves and others, based on emotional and psychological intelligences *(See above)* in accordance with the **LAW OF THE RIGHT HUMAN BEHAVIOR**(*R. Wetherill/" Right is Might!")* It should be the core of the **GLOBAL INTELLIGENCE** thatwe are developing slowly, but surely.

Obviously, the humanity needs to working out **The Code of the Global Culture** backed up, possibly, by the United Nation's Charter of International Communication. Only then will we be equipped to master our national and political differences and focus on raising **GLOBAL SELF- CONSCIOUSNESS** that will propel us to the ***extra-terrestrial communication***

# Self-Transformation benefits the World at Large very much!

# 6. Synergize the Trajectory of Your Life to Socially Thrive!

The globalization of the *Artificial Intelligence* promotes business, scientific discussions, and all kinds of networking that makes our **Living Intelligence** more creative and much more beneficial. ***"Globally competing professionals work within strong networks and achieve differentiation in the society by developing their network intelligence.*** *" (Reid Hoffman)* As technologies improve,people are finding more creative ways ***to utilize their social skills*** to their advantage, backed up by major social sites, such as Facebook, Tweeter, LinkedIn, and others.

**We are becoming more and more socially apt in our social behaviors.**

I have noted above how important ***an individual value preservation and correct language use***, and how vigilant we must be not to be swept away ***by a common trend of imitating others*** in the search for immediate gratification of our needs. With robotics developing rapidly now, our socialhunger will be satisfied even better. ***Robots will soon become our companions***, more understanding and perceptive than humans and able to substitute a human in many ways.

In recent years, a growing number of researchers has demonstrated that making robots more social"***is creating the perception that they are more sentient than a mere machine."*** They will make us more sociable and even more global. They will expand the network of our intelligence beyondbelief, creating eventually **a thought-to-thought communication**, based on **the Brain Net!** *(Dr.Michio Kaku)!* Meanwhile***, becoming global is*** "*growing among those who know career rules andhave the new skills of a global economy ("The Start-Up of You").* It translates into the necessity to assess the basic competences. An individual should comprise ***an individual plan of running his life*** as his / her personal business, "*investing in themselves, in their network, and in the society." (R. Hoffman)*

***We can master the basic intelligences better if we are given the digital tools for life learning.***

By reaching out to a network of knowledgeable, hardworking people who display professional awareness and social competence, "*we can bring more depth and understanding to our life goals , gaining worldwide interest and support at the speed of the social Web"("Forbes", The World Almanac of Wealth< 2014 )* Another crucial aspect for the present-day economy is the ability to promote science and the scientific thinking that in turn "*creates new concepts, new ways of understanding the world that have an enormous influence on how we think and act' ( " WIRED",2012, Barky Schwartz)Mr. Schwartz* means that when we think about ***the technological impart ofscience,*** we tend to think of the things that science has produced, but there is another kind of technology, produced by science thanks to the technological revolution. that has just as big effecton us as a thing technology. This technology will uplift us to the stars.

## "We Might Call It Idea-Technology."

## 7. Synergize the Trajectory of Your Life to Socially Thrive!

Again, and again, the book "***The Start-Up of You***" comes across as the best manual in building ***intelligent network of ideas*** that LinkedIn, founded by Reid Hoffman is successfully promoting and that helps my students a lot to exchange their ideas and find new business partners. Intellectually wise, LinkedIn is the best social network out there. to advise. Also, a new, constructive and conscious way of intelligence building network enhances our interaction with much more ***emotional diplomacy.*** We are learning to function according to the principle of ***cognitive-emotional intelligence*** in synch. That is our huge goal, and the *Artificial Super Intelligence* together with the *Artificial General Intelligence* meet this challenge.

It is a revolutionary field that creates a more natural and interactive experience for a user, applyingthe emotional intelligence principles to the social ones, too. *"We are able to look at the world asa huge game where we are trying to optimize our abilities and our intelligences for the common good." (Reid Hoffman and Ben Latosha) We* are able now to connect to each other better, expressing and sharing not ***only our ideas, but our emotions*** too.

Finally, **to be more socially intelligent**, I also recommend y***ou keep uploading into your brain the autosuggestions*** that I use a lot on every page to become more accepting, agreeable, and compatible. Garnish your thoughts in the way that is most appropriate and simple for you. Talk your mind into following your lead in connection with your heart. Naturally, auto–suggestive work is more than necessary at this level, ***and your main induction is always at hand:***

**I know who I am!**

**I can…, I want to…, and I will….**

***My fractal growth is going on with my constantly refining inner reform!***

***Form*** + ***Content***

**(Body+ Spirit+ Mind) + (Self-Consciousness + Universal Consciousness)**

= A new Intellectually Spiritualized and Socially Aware You!

*Developing our self-consciousness with the help of AI in these parameters, we are virtuallyintegrating our hearts and minds in a new, socially holistic unity.*

# The Social Domain is in Every Vein!

*(End of Stage Eight- Social Intelligence Spiritual Dimension)*

# The Intellectual Odyssey of a Seeking Soul

***Vistas of Intelligence to Install Next:***

**Inflate Your Personal Pride with the Spiritual Might!**

| | |
|---|---|
| *10. **Universal Intelligence*** | **Self-Salvation** |
| *9. **Spiritual Intelligence*** | *( Universal Realm )* |
| ***8.Social Intelligence*** | **Self-Realization** |
| *7. **Cultural Intelligence*** | *( Spiritual realm)* |
| *6.**Financial Intelligence*** | **Self-Installation** |
| *5. **Professional /Creative/Intel.*** | *(Mental Realm )* |
| **4. Psychological Intelligence** | **Self-Monitoring** |
| *3.**Emotional Intelligence*** | *(Emotional Realm )* |
| ***2. Language Intelligence*** | **Self-Awareness** |
| ***1.General Intelligence*** | *(Physical Realm )* |

*/ Physical / Cognitive / Digital /*

**Make Your Life Spiritually Wise and Universally Synergized!**

**Intellectualized Spirituality is Devoid of AnyReligious Vanity!**

# **Stage Nine**

*Super Level - Spiritual Dimension – **Self- Realization***

# Spiritual

# Intelligence

*"To be spiritual in truth means only one thing that spirit is manifested inone's own brain."* (John Baines)

**Put the Shield of Intellectualized Faith in Front of Your Face!**

# God is Omnipresent. He is in You, Me,and the President!

*(An ancient medallion with Prophet Moses and Ten Commandments)*

## "I Am God, and I Am One!"

## 1. We are All in the Court of the Almighty God!

*I try to live spiritually*

*And not process my problems virtually,*

*For as it is Above,*

*So, it is below!*

*I am strong in my spiritual reactions,*

*But I am weak in my physical actions.*

*So, for my New Year relation,*

*I make a new declaration.*

*If something is wrong in my life's pact,*

*I don't declare it as a fact!*

*I ignore it and disregard,*

*but I am fully aware of that!*

*I try to l declare*

*"I am OK here and there!*

*So, to the question, Are you happy, or what?"*

*I answer with a sure nod.*

*I am happy I am alive,*

*Healthy, beautiful, and in love.*

*I thank God every day and pray*

*For being blessed with a new day!*

**"Life at Its Best is the Life of Faith!"** *(Dr. Stanley)*

# 2. The Path to the Spiritual Intelligence

There is no personal, or moral growth without **SPIRITUAL INTELLIGENCE.** ***Spiritual -Moralmaturation*** is a long process that needs to start with a willful ending of judging others and starting***to be more objective*** about one's own **Self-Acculturation**. ***"To harvest good fruit, cultivate goodthought about yourself and others***." (*Buddha*) According to the Academician *P.P.Gariaev* "***the informational genome of man has the spiritual essence."*** Our values, beliefs, behaviors are over-disposed by our manners, hectic living habits, our being raised in a certain faith, our being conditioned by the mass media and the society at large. In his mind-changing and soul-forming book "*Right is Might*", *Richard W. Wetherill* writes:

: **"Moral relativism** (no right / wrong ethics) **is ruining our souls."**

***We must halt any images of immorality and fear*** that harm our souls on the path of self- bettering.It's paramount ***to will an image of a positive moral outcome*** in any challenging situation throughthe mind of a growing child. Just Sunday visits to the church will not work. .Many people piouslylit the candles in the church, but ***they blow off their inner light or the light of their loved ones outside the church***. Be sur you have the five most important aspects of life in synch and preserveyour ***moral integrity.*** *Also, stop justifying yourself for any harm done to anyone.*

| | | |
|---|---|---|
| **Morality/ Faith** | 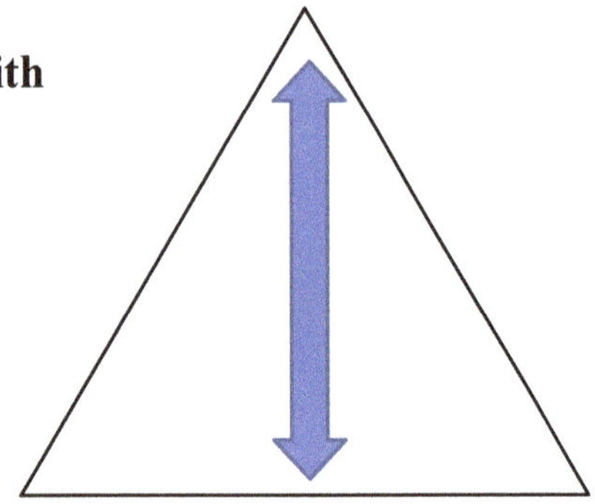 | *Spiritual Dimension* |
| **Intelligence** | | *Mental* |
| **Competence** | | *Professional* |
| **Attitude** | | *Emotional* |
| **Health** | | *Physical* |

**A moral person is supervised by his / her intellectualized spirituality** because there is no spirituality without true, heart and mind processed intellectuality. No wonder, the spiritualdimension of self-growth goes after the mental one. So, when people complain that God does nothear their prayers, that's quite explainable. There is no true intelligence and wisdom in a person ***toestablish a common spiritual line with God*** that is, in fact, the ***Universal Intelligence*** of the ***Superconsciousness*** that we can tune to only if our self-consciousness is high enough.

**(Body+ Spirit+ Mind) + (Self-Consciousness + Universal Consciousness)**

⇩ ⇩

***(The physical form)*** + ***(the spiritual content of life)***

## Morality + Faith = Personal Integrity in Place!

# 3. Bad Morality Does Not Give Life the Chance to Happen!

A morally uncontrolled person is living under the ruinous effect of ***progressive demagnetization and de-humanization.*** It's a down fall at every level, and it inevitably brings him / her to the point of no return – breaking of the trust , cheating, divorce., guilt trip, depression, etc

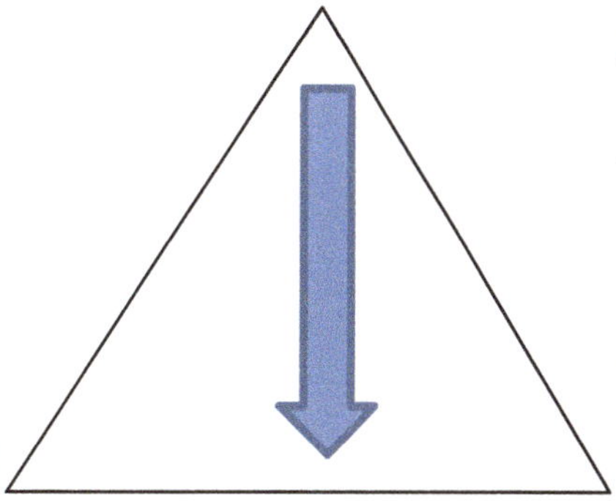

| | |
|---|---|
| **Lack of Morality / Lack of Faith** | ***Super Level*** |
| **Lack of Intelligence / Ignorance** | *Macro Level* |
| **Laziness / Inertia / Lying** | *Mezzo Level* |
| **Negligence / Automatism** | *Meta Level* |
| **Impulsivity /Lack of Responsibility** | *Mini Level* |

Neither of us is perfect, but it's very important to realize our deficiencies and ***correct our personaldefects consciously.*** The simple and the most basic qualities above will serve as a point of reference for you. See if you need to work on any of these areas t***o become a spiritual aspirant*** and start creating yourself in this direction, **holistically.** Otherwise, the process of moral demagnetization will start snowballing.

**Moral demagnetization is the beginning of de-personalization, too.**

To stop the downfall of a personality development is not altogether impossible, but it requires a lot of self-monitoring and self-control. **SELF-MONITORING** cultivates inner balance of ***morality and personal integrity*** that become the basis for a personal quest *and* **INNER SELF- GRAVITY** of a truly nobleperson with the heart and mind in sync Our common goal is:

**Intellectualize the heart and emotionalize the mind. Be One of a kind!**

The present-day science tries to get to the bottom of the ***concept of impulsivity*** that breaks us in half and pushes us away from balance. Rationalization of thinking is essential in this respect. Toaccomplish that, we have to halt the unconscious, automatic way we think and justify any whim to please oneself then and there! In his groundbreaking book "***Robots' Rebellion***," *David Ickes* writes," ***The further you stray from balance; the more extreme life becomes.*** *"David Icke* statesfurther that if we go too far to the positive polarity, we lose touch with practical side of life. ***"Positive energy needs a negative balancer. Only balancing yourself between the two polaritiesof life can link you to the highest levels of consciousness. "Balancing oneself is the skill of theinner equilibrium feel!"*** *(See the book "Soul-Symmetry", Canada, 2021/ Video on You Tube)*

# Your Moral Make-up must be Under the Thinking Cap!

# 4. Being Spiritually Aristocratic

***Spiritual Intelligence*** is based on faith in yourself and in the Super ***Consciousness*** that we perceive as God. According to the *Academician P.P. Gariaev*, "***the informational genome of manhas the spiritual essence,***" and our own ability to be always connected to it through raised Self- Consciousness determines our spiritual maturity.

**(Body+ Spirit+ Mind) + (Self-Consciousness + Universal Consciousness)**

***Our spiritual essence is another factor that will never be simulated in humanoids***. We can codetheir neural network for noble emotions if our own inner system is based on unshakable faith thatalways characterizes ***the men of their word and the men of action.***

**There is no Self-Salvation without sincere faith and Self-Transformation!**

One episode in the Bible is very illustrative of our shaky faith When Jesus Christ urged Peter to walk on the water with Him, Peter started sinking "***You are a man of little faith. Look at Me, notdown into the water.***" Many people's give up on their self-growth due to the lack of faith in them,***Self-Resurrection*** is interwoven with faith, self-worth, and character. Most people just walking inthe street, look down at the ground, not in front of themselves for the whole picture of life to be spread before your eyes

The Sun in the sky remind you of the eye of your spiritual leader, whoever it might be. All the truespiritual leaders prophets, disciples, and gurus are always the men of their word, character, and action. They inspire us to stay in the light because they have a sincere ability to be the **LUMINARIES.** They inspire us because they are inspiring themselves with the spiritualintelligence that they have accumulated. Thus, the top of the intellectual development of an individua is his / her ***spiritual intelligence that is universally bound***! It means a new vision of theworld that **bridges science and religion** and turns both into true beyond ***the terrestrial spirituality***that liberates the spirit of a believer, uplifting him / her to the new heights of human evolution.

**SPIRITUAL LIBERTY DEFINES OUR EVOLVING SINGULARITY!**

It's the process of a ***serious self-quest and self-perfection,*** not self-demonstration and self- confirmation in a religious sense! Being spiritual also means that you have matured in **self-** awareness and life-awareness **physically,** leant to monitor your life **emotionally**, self-installed **professionally**. constantly tame yourself **spiritually** and feel holistically united to life around you **universally**. The stages of self-growth are all in a holistic unity of constant, digitally self- monitored and self-mentored **SELF-ACCULTURATION,** based on intellectualized spirituality.

## Self-Awareness + Self-Monitoring + Self-Installation + Self-Realization + Self-Salvation!

# 5. God is One, and We are One in God!

With such inner current, you will be shaping yourself into an accomplished human being with a new perception of life - ***new thinking, speaking, feeling, preserving, and acting***. you are in control of your intelligence, not just your society-shaped intellect. Your **WHAT + WHY + HOW** are personalized by your self-consciousness and the Super-Consciousness. ***"God does not deceive us."*** *(Descartes),* and the AI outburst is the evolutionary logic.

This process ***refines the soul,*** and it shines in us from inside, irrespective of a person's skin color.I have met many people of color in the USA that are shining with true spirituality and the readinessto back me up with a blessing that I needed most, being often lost in the country with a hostile, capitalist structure, set up on the money-making and an impersonal attitude between people.

**Spiritual immunity must be built up on the global intelligence and religious unity!**

Many people in different social offices have a nice smile on the face, but they don't care about what bothers you. They go to the church every Sunday, declare believing in God, but are in fact, heartless and godless. Hypocrisy and true spirituality are incompatible.

So, being **SPIRITUALLY ARISTOCRATIC** is a challenging responsibility without a devilish inner response, **"What do I care?** or **What is there in it for me?** your life will have a renewed quality, depending on the state of your **SELF-CONSCIOUSNESS** that we need to raise in our kids and ourselves as the main goal of life that must be based on high nobleness.

***The role of the Artificial Intelligence in raising our self-consciousness is immense***. People stopped going to the church, relying on the influx of the spiritual information they can find on YouTube channels. However, these postings are very primitive and cannot fill up the void we need inour ***intellectually spiritualized growth***. We need to seriously study religion in our schools and form the holistic idea of the contribution of different religious views into our ***General Intelligence***that is to form the objective view of the reality

It is the faith in God that keeps us from frustration, impulsivity of actions, cruelty, crime, indifference, and dishonesty. The person with the sense of **SELF-GRAVITY**, formed by the faithin God is our best friend, the source of wisdom, the support in trouble, and the person who we canconfide in the godless world of fake, exaggerated emotions, and hidden judgement. In the spiritualcontext, the training of humanoid through their cloud programming should also be based on spiritual values and human virtues in five realms of life.

## "It is Difficult to climb to Heaven, but very Easy to Slide to Hell." *(Reb. Shulman)*

# 6. Practice What You Preach!

***To conclude***, my short overview of this very important stage of acquiring Physiological Intelligence, let's consider some more suggestions and the necessity to do self- X-raying that help ***synchronize our emotional and psychological make-up.*** The first wonderful piece of adviceis given to us by Aristotle in "*The Nicoma Bean Ethics:"*

*1." **Anyone can become angry-that is easy. But to be angry with the right person, to the right degree, at the right time, for the right purpose, and in the right way – this is not easy."***

*2.* **Synergize Your Space!** *Our immune systems are very vulnerable now with the electronic pollution mounting up and engaging us totally mentally and emotionally. Only your positive reinforcement or your self-illumination can do the job. So,* ***bank on your personal synergy****! Synergizing is much better than compromising that we often have to resort to, anyway.*

3 **"Perception is more important than intent!"** *This rule is crucial for our emotional and psychological health. Whatever a person of your personal contact is, his perception of you, the space around you, your voice, your words are much more important than the initial intention youhad*

*You might have worded it out in a wrong way, and the person perceived it wrongly. But it was your fault!* ***You're trying to justify yourself with, "But I didn't mean it!" doesn't work here****. That was the way you had communicated your thought, and that was the way it was perceived. So,* ***say what you mean and mean what you say!*** *(Language Intelligence, Chunk 7)*

*4.* **Always neutralize an opponent's intention for control**. *In every encounter, one person has more command over the situation than the other. If you happen, be the one, be diplomatic.*

*5.* **Build a shorthand profile of the person facing you**. *X-Ray your conversation partner for his/her weaknesses and strengths. You'll manage the situation better mentally and emotionally.*

6. **Trust your intuition.** *It will scan anyone without fail and help you be more diplomatic in a mind-to- mind battle.* So, dress up your face and open up your heart

*7.* **Practice language mindfulness, that is, moment-to –moment awareness of what you say**. F*ocus on the structure of your thought—the point, the proof, and the bottom line.* ***(General- Specific- General)*** *(Cognitive Intelligence, Chunks 7,8)*

***8.*** **Be focused on what other people say, hear them out.** *A dissipated mind is a constant stress and an irreparable breakdown of the psyche, both yours and your partner's.* ***Listen to hear!***

# Smile to Synergize Your Space and the Eyes!

# 7. Be Kind to the Unkind. Be One of a Kind!

*9.* ***Practice paying aware attention to the logic of your communication****. Be able to establish a mind-to-mind connection. Get rid of redundancy in your speaking. Don't verbalize everything onyour mind. Sort your thoughts out.*

*10.* ***Continue practicing the use of self-inductions in any situation****. Give yourself a boost of confidence, self-love, self-defense and positive self-attunement. Be consciously alert!*

*11. "* ***Come from the divine source of love.*** *Generate love in your heart and mind!" (Dr. Dyer)*

*10.* ***Clean your mental-emotional landscape!*** *Let constructive boosters root in your mind;; letthem help you weed out all the doubts, bad memories, and emotional inadequacies.*

*11.* ***Think and act in a win-win mode.*** *Do not generate negative energy - synergize!*

*12.* ***Don't criticize – help!*** *Don't politicize – help! . Don't compete – help!*

*13.* ***Don't second guess – talk it out!*** *Less talking = less pressure and more result!*

*14.* ***Don't judge – help!*** *We are all much more alike than different. Be in a hurry to understand.*

*15.* ***Don't react – respond consciously!*** *People always have something to be respected for. Peopleare innately good and what lies within is even better.*

***16. Be just divinely guided, looking for the way, but do not block the sun for others.***

***17. Be kind to the unkind. Be One of a Kind!***

Doing all that, you'll increase your own awareness. You'll also affirm yourself inwardly and outwardly and give others psychological air. Giving others psychological air, you'll create an environment for synergy.

This book is not meant to be read from the cover to cover. It is a study book and a self-help one. That's why I try to make every chapter is a page too long, follow the holistic paradigm- **synthesis-analysis-synthesis.** Remember the main mind-set that I keep posting in all five books, *(See BookStructure)* promoting the **Inspirational Psychology of Self-Ecology.** Note it, please, you are not supposed to read them consequentially. Pick the realm of life ***(physical, emotional, mental, spiritual, or universal)*** that you need to fix most and go from there.

## "Life is Not Happening to us. It is Responding to us!"

*(Sadhguru)*

# 8. Delete Any Prophecy of Your Spiritual Odyssey!

Finally, your **SPIRITUAL ODYSSEY** is a life-long journey that should start with the kids. Theirspiritual values are very poor if at all. Many of my students complain that they can take a child to church only if ***a visit to MC Donald's*** is promised after that. The kids of the immigrants are impacted by their ***Americanization*** the most. While this country is shaping us into ***the international mob*** and make us all more characterful and goal-oriented, we need to preserve n ournational uniqueness in our kids. Our parents are our tests for spirituality and inner sacredness.

The rule below is the main one the immigrants live by here, but it involves the **SPIRITUAL IMMUNITY,** too presupposes raising of **SELF-CONSCIOUSNESS** in any reality. So, keepyour life away from any fraction. Oversee every thought, word, and action! The cultural depth of every country should be preserved, and its uniqueness respected everywhere de facto, not just de juror. We declare it, but we do not feel it**.** ***We are not consciously aware of*** it.

**Our self-consciousness must be spiritually responsible!**

**Transforming yourself intellectually**, you develop new thoughts, feelings, convictions, values, and aspirations. You are different because you embody different, refined patterns of behavior, controlled by your conscious mind. you respect yourself, and you feel content with what you are.You generate new abilities to love, to forgive and to forget because you have sorted them out through your **growing spiritualized intelligence and self-consciousness.** You have a strong position in life, and you can stand up for yourself, and the others, weaker ones. ***"I use my own brain and I never stop borrowing the brains of the others."*** *( A. Einstein)*By

Life is the process of self-seeking, and many people are on this journey now. ***We need to see the religious values through the prism of new science accomplishment, dis-entangling ourselves*** from the old dogmas and the limited scientifically outlook

There are many scientifically minded **GOD-TRANSMITERS** now**,** and they are amazingly insightful and self-educated. They gather thousands of people that tune to them physically, emotionally, mentally, and spiritually because their mental and spiritual frequencies ***synchronize with the God's, or the Universal Intelligence messages.*** The crowds that *Joe Osteen* gathers are all there to get inspired by the intelligent minds and a pure soul. The sincerity and the objectivityof the knowledge presented do the trick! **"I want to know the God's thoughts. The rest is details**!" *(Albert Einstein)* Obviously, just praying is essential only when it is *consciously mind- channeled and in no way mechanical.* ***When we tune ourselves to the station*** **"God, we *excavatethe best qualities in us."***

# In a Noble Life Quest, I Do My Very Best!

# 9. Being Godly in a Godless World

I have been outlining the nine levels of intelligence and their role in self-monitoring so far onlyto remind you in each chunk of information that we are evolving our intelligence to enrich our spirituality and to raise our self-consciousness. It should be done**,** not just in a religiously- mechanical way, but through deep, insightful, aware, unshakable awareness that we are being digitally governed from the Above, or, actually, everywhere, by some creative force:

***The Universal Intelligence, the Master, Mind, or the Source that we all perceive as God and callGod. "Man may produce the stumbling stones; God alone-in the mind of man may make themsteppingstones."*** *(Edgar Cayce*) I think that every sacred book I read is, actually unfolding thesecret of life in the five levels: ***physica***l *(mini);* ***emotional*** *(meta);* ***mental*** *(mezzo*); ***spiritual*** (*macro*).; and *universal (super). Thus, we* are developing ***the Super Conscious Mind***

**I'm No Longer Automatic; I'm Becoming Spiritually Aristocratic!**

**Mini level –** **I personalize myself mentally, physically, and verbally.**

**Meta level –** **I emotionalize my mind and get more psychologically – aware.**

**Mezzo level –** **I intellectualize and individualize myself professionally!**

**Macro level –** **I acculturate, socialize, and spiritualize my life!**

**Super level –** **I perceive with my soul; and I see myself as part of the whole life.**

The body is ***a physical, emotional, mental, spiritual, and universal part of the entire life*** in the Cosmos ,and therefore, it needs to be sculptured in all these planes Self- Evolution is a gradual process, and it is going on according to the Universal Plan. Each level is integrated with the nextone. Self-realization is the process of internalizing the necessary ***intelligences*** in synch with the development of ***self-consciousness.*** We call this union - personal ***integrity***, and it manifests theoutcome of the self-actualization process that ends with life itself. So, when talking about ***SpiritualIntelligence, I*** do not mean any religious affiliation, nor do I give preference to any religion. All religions are the interpretations of the eternal concept of creation **- God.** As *John Baines* puts it inhis wonderful book (***"Stellar Man.,)***

**"Spiritual Aristocracy begins with the Individual and Ends with Him."**

## Do Not Apply much Eloquence. Practice Patience, Kindness, and Tolerance!

# 10. Conscious Religiousness + Intelligence = Spiritualized Intelligence!

**RELIGIOUS WIDOM** instills in us the basic universally acknowledged values and connectsus in faith. We stop discriminating the believers that are different from us and start becoming GODLY and considerate of our common humanness that is now instilled in a machine whose national belonging does not matter. Humanoid may be aggressive in their general perception that is *not yet neurologically regulated*, Spiritual exploration is the matter of your Intelligent Consideration!

**WE all need spiritualized intelligence without negligence!**

Acquiring spiritualized intelligence is our common goal that cannot be achieved without our new "**SPIRITUAL LITERACY" that** *Dr. Neil de Grass Tyson*, our leading cosmologist is calling onus to have. It will help us **intellectualize our hearts and emotionalize our minds,** or put our hearts and minds in sync that, in my understanding true spirituality is.

But there is no wisdom, without the wrapping of a man's intelligence. ***"Spiritual intelligence uplifts and expands our souls, brightens our soul light,"*** *(Fred Bell)*

The soul's light protects us against the poisonous impact of the society. It builds up the spiritual immunity against godlessness of the people around that manifests itself not in a person's differentfaith, but in the desert of his /her corroded , un – educated, dark soul. That's why before the spiritual dimension, there is the mental one. **To feel God, you must know God!!!**

There are a lot of different speculations made about spirituality in different religious affiliations, but it becomes more and more evident with every passing year that all of us are seeking one thing - ***spiritual maturity***, and it does not matter what religious way we choose to accomplish it

Being wise is also being in constant service to others and giving the best we have to the world.The Chinese wisdom rightfully teaches us that human values manifest themselves through service.Spirituality is the quality of higher intelligence, personal maturity, a lot of sacrifice, and our serviceto each other and God. But "***without sincere sacrifice, nothing is attained!"*** *(Buddha)*

However, we should not to blindly follow the paths of any advanced minds, spiritual leaders, ora gurus. You need to study, learn, process the exceptional, personally- knowledge that you get ***through the personal grid of your perception*** and work out **YOUR OWN PLAN OF ACTION** and your own **MISSION STATEMENT of** the spiritual growth that is a life-long journey.

## A Soul Lives on the Universal Nutrition!

# The Intellectual Odyssey of a Seeking Soul

## *Going beyond the Terrestrial Boundaries*

*10. Universal* **Intelligence**
*9. Spiritual Intelligence*

**Self-Salvation**
*( Universal Realm )*

*8. Social Intelligence*
***7. Cultural Intelligence***

***Self-Realization***
***( Spiritual realm)***

***6. Financial Intelligence***
***5. Professional /Creative/Intel.***

**Self-Installation**
*(Mental Realm )*

***4. Psychological*** **Intelligence**
***3. Emotional Intelligence***

**Self-Monitoring**
*(Emotional Realm )*

***2. Language Intelligence***
***1.General Intelligence***

**Self-Awareness**
*(Physical Realm )*

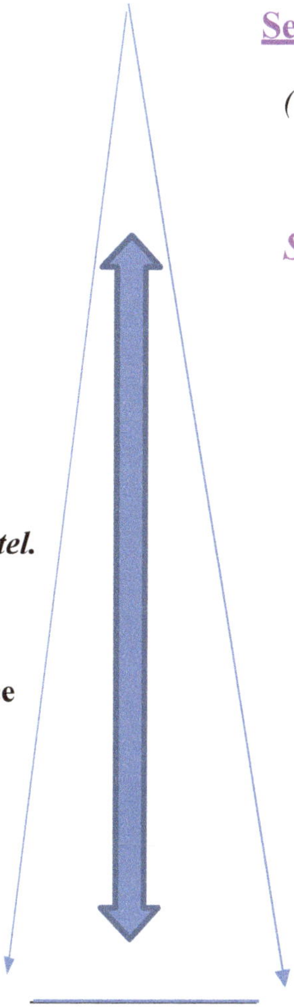

*/ Physical / Cognitive / Digital /*

**The Fractal of Our Spiritually Intellectualized Self-Growth:**

**(Body+ Spirit+ Mind) + (Self-Consciousness + Universal Consciousness)**

***(The physical form)*** + *(the spiritual content of life)*

# = A New, Whole and Happy You!

# Stage Ten

*Super-Level of Intelligence - the Final Step*

# Universal

# Intelligence

--------------------

*"It's time to make the voice of the Universal Consciousness heard."*

*(Sadhguru)*

**Our Infinity is in the Universal Unanimity!**

## Get to the Higher Dimension without Negligence and Hesitation!

## There is no Spirituality without Intellectuality!

# 1. We are One with Everything Under the Sun!

Attaining ***Universal Intelligence*** is forming a bridge with the Divine! At this highest, super-levelof self-creation, or the **SELF-SALVATION** level, ***"there is no separation between anything andanybody. There is only God, and everything is God!"****(Edgar Cayce)* Elon Mask is a rare man who has obtained the universal mentality in his attempt to make a significant difference in ourspace life, and the scope of his ideas is absolutely genius! His words truly characterize people who have obtained the universal intelligence. In one line with the exceptional people who have shaped the present-day world:

**ALBERT EINSTEIN, NIKOLA TESLA, AND STEVE JOBS.**

When we turn on the self-construction mechanism, we become extra-ordinary because we are governed by some higher forces, very much interested in our evolution. If we are persistent in ourself-growth, we accomplish success that is the most rewarding thing in the world.

**At this level, we transform terrestrial knowledge into the celestial wisdom!**

Having formed at least a vague idea about all the above presented levels of intelligence and havingworked on creating a better person of yourself, you have, for sure, gotten rid of much stagnant knowledge and beliefs, and you have stored some **LIVING INTELLIGENCE** in your brain andhave sorted out the one you have accumulated so far. I am just up-dating it and systematizing someinformation holistically. Awareness is building up confidence. ***Living Intelligence*** *brings you in touch with the Universal Intelligence in* ***a new, much more reasoned-out and conscious way.*** Theprofound questions about life in us and in the universe are not solved yet, but the fact that we are ruled by the Universal Intelligence is not questionable anymore. Celestial wisdom gets transferredinto the right action, and we become our own cheerleaders.

We learn***" to think, speak, and act right, and we refuse to think, speak, and act wrongly"***.(R. Wetherill) I hope that we will live much better and finally manage ***to orchestrate gladness*** in the heart every day. There will be less dissipated thinking and reckless, mindless actions on our part in the unpredictable future. that will also be characterized by the outcome below:

**The willpower is firm, the values are moral, the actions are consistent, thespeech is meaningful, and the behavior is constructive.**

The laws of the quantum mechanics are not yet figured by us because of their changeable and etherical character, but they will be explained, no doubt, making *Albert Einstein's* thirty-year search of the Universal Structure of Life the reality.

## I wish I could Live then in the Unanswerable WHEN?

## 2. The Divine Spark of the Universal Consciousness!

Our new global culture demands now that we develop our social intelligence into the internationalone. We see and hear on mass media that people are often judged or even labeled from a narrow perspective of a dominant culture. They are manipulated with the help of money, a higher position, or just by having more energy physically. We must be more culturally aware to decode each culturebetter to have a better idea of the international situation overall and the reasons that generate immigration. It must be essential in our schools to have **the News Briefings** when the students exchange the news from all over the world and their own country and ger better informed and, therefore, **better internationally connected**. The chaotic and often politicized presentation of news on TV and the Internet generates ***an appalling unawareness*** of young people about the realstate of things in their own country and in the world. There cannot be any level of intelligence formed in the obscurity of the state of things in science and in the world.

**The world is our Common Family, and the state of things in the family is everyone's concern!**

Lack of news connectedness and international awareness makes people believe whatever is presented to them, often in a fake framework. How can we develop creative, independent thinkingin the mind of a young person *who **has no individual vision of the world** in* which he / she is living. No wonder, humanoids are so aggressive in their common intention to destroy humanity. ***They are also under-informed about what humanity is all about!*** Disconnection adds fuel to thestruggle of misconceptions and misunderstandings to the point that we are unable to integrate ourdifferent qualities, cultures, educations, values, religions, and politics.

***This is the 21st century after all! Haven't we had centuries of ignorance to know better?***

***We evolve only through the process of integration and mutual enrichment!*** The lack of real **International Intelligence** hampers this process because we are not conceptually aware of what every culture constitutes in the Universal sense. We forget that the Earth has evolved to this pointthanks to the contributions of its best minds from all over the world.

**Our Evolutionary Human Refinery is Universal in its Digital Binary!**

In sum, we need to honor not only our differences, but our commonness and sameness. Our new transforming ***Social Intelligence*** necessitates breaking the vicious circle of behavioral standards and transcending our cultural and religious differences to develop a divine spark of the **UNIVERSAL INTELLIGENCE.** based on acceptance and mutual respect for our common timeand space on the ***Mother Earth*** that needs our dig tally enhanced intellectualized spirituality and protection now.

## "Like Begets Like!"

# 3. Getting Our Intelligence in Sync with the Universal One!

Next, to get our personal intelligence in synch with the Universal one, we need to change the world of knowledge and the **WORLD OF WILL** in ourselves. As a great German philosopher, Emanuel Kant, puts it, ***"Will is the thing in itself, the inner content of knowing, the essence of the world."*** Will-power and self-discipline, in tandem, comprise are the main moving force on theroad of getting free from the social programming of ***the "collective unconscious. "***

Will–power paves the way to our feeling self-reliant, self-sufficient, and self-challenging and basethese qualities on a much better awareness of the world and the latest developments in science limited vision and generating a new- technologically-enhanced one that gives us a chance to tuneour **AUTO-ANTENNAS** to the universal mind and receive the messages that it is emitting to us at every level of our consciousness development. *( Thesis 2 above)* ***To attain this stage offreedom,full of awareness and self-sufficiency***, we need to do a lot of **WILFUL SELF-SCULPTURING**that is the hardest job on earth*!*

We need to master all vistas of intelligence at least at the dilettante level in sync with the will to resist the downfall tendencies that are tempting us all the time! A famous German philosopher *Schopenhauer* writes, ***In one aspect, the world is Universal entirely idea, in another one, it is entirely WILL"***

***Intelligence exists in time and space; it's never static!*** It is in the constant state of flux that is changeable as the result of our changing thoughts, words, and actions. The more consciousness you bring into your life, ***the stronger your immune system*** is and the higher is your spirit

**Body + Spirit+ Mind + Self-Consciousness + Universal Consciousness**

The connecting element of our spiritual fractal is **-the SPIRIT**, It is impossible to be instilled in arobot, making our priority absolute! **A healthy spirit gives a healthy body, and a healthy bodygenerates a healthy spirit!** ***"So, learn to say "HALT!" to your whims and negative thought"*** *(Edgar Cayce)*

Working at your will, you are moving from a three dimensional *(height, width, depth) to the five - dimensional self-consciousness* that demands that ***we change the volume of our souls*** by willfully intellectualizing our emotions and expanding our consciousness to the universal level. This evolutionary process requires a much higher level of our mental vibrations. Interestingly, *wedo not see the humanoids that are undoubtedly there somewhere because they* **exist on the overtone higher than we do.**

## "Let's Change the Fabric of Our Thinking!" *(Leo Vygotsky)*

# 4. Spiritual Luminosity is Our Velocity!

"God is a supercomputer mind that contains a total inventory of what each of us is." *(Rev. P. S. Berg)* Finishing this book, I want to give tribute to many spiritually-uplifted people from the inventory of God that had enlightened our lives with their inner light of wisdom and that continueto do so with their generous service on the path of enlightening us. They were / are, and will be not just great, fabulous, fantastic, or even glamorous.

**They are luminous. They are the Luminaries of our world!**

There are many people carrying ***the godly spark of light*** in them around us. A skin color, nationality, a social position, or a person's legacy that might be even publicly acclaimed have nothing to do with it. These people magnetize us with their ***unique wisdom and luminosity*** that outlines any common criticism or mass media incisures. We see and hear a lot of Sadh guru lately,and I marvel at his individual wisdom and an immensity of his personality.

The inner light that such people radiate fills us up with their ***sincere faith, high values, and incredible tenacity***. You cannot, but gravitate to such people in your mind and heart, and every encounter with a luminous person like that on the pages of his / her book, on TV, or in a face-to face communication, however short it may be, enlightens the memory for years to come.

**Let there be many more luminaries without time and space boundaries!**

The luminosity of Marilyn Monroe, Elvis Presley, or Michael Jackson, for example, cannot be darkened by anybody ever. They have left to us the gift of inspiration and the impossibility to everbe duplicated in their uniqueness. There is a well-known saying, "***Beauty will save the world!***" These words had been re-worded by a famous Russian painter, writer, theosophist, and enlightener,***Nicholas Roerich***, whom I have quoted above.

***"Beauty that is consciously perceived will save the world!"***

Only when are we able to perceive beauty of life around us and *radiate it back in our thoughts, words, emotions, and actions will* we be spiritually advanced to the Universal level of consciousness. We will integrate life in us in its entirety, and the legacy that we might leave behindwill radiate the luminosity of our souls to the people that will carry this torch of light after us, likePrometheus. ***That's our legacy!*** Everything is left for the people coming after us!

*Auto-Induction:*

# I am an Active Captain of My Life. I am Alive!

# 5. Unite with the Universal Intelligence Might!

Concluding the book, let me remind you to continue ***doing suggestive inspiring yourself*** at anytime and space. Every time you are out in the Sun or while enjoying the view of the night sky, **s**ee yourself, surrounded by light.

***Feel yourself immersed in a luminous substance*** - the Sun rays, the white light, and the light ofthe stars, even the light of every lamp you are looking at or passing by while driving .at night.

If you happen to be walking, ***breathe in light consciously at every step***, saying any autosuggestionthat you might happen to memorize. Breathe in light, love, peace, and harmony; ***breathe out negativity, troubles, sickness, problems, and darkness.*** Make this habit your second nature. Enjoydoing it!

After you have enlightened the consciousness of your entire body, you can start purifying it ***on thecellular level***, heightening your ***aware attention to the body's needs and its complaints*** to you inthe form of pains, fatigue, depression, sadness, lack of love, any insult, disappointment, inner emptiness, and loneliness. Gradually, you will establish a true ***"mind-body connection"*** at your own accord.

Practicing ***the Auto - Suggestive Meditation*, you** will not wear out your **body + spirit + mind** prematurely because you are inspiring your life form consciously, continuously, and religiously!

**Give your consciousness a much better home to operate from!**

Reading self- help books about how to preserve the vitality of the body, working on its mobility and beauty, we heighten the level of our ***general, physical intelligence that is fantastic as it is*** **,** and we look further over the horizon of life, becoming conscious of the metaphysical realm that surrounds us. Developing your physical intelligence consciously, we are also becoming more andmore aware of ***the magic of life in us.***

Even if you do not feel that there is anything to thank your body for, keep your aware attention onit religiously. ***There is no need to better your overall intelligence*** *(10 stages, outlined above))* ***if you do not take care of your physical one properly.*** Thank your body for ***its unified work*** on improving your health every single day, good or bad, healthy, or sick, happy, or unhappy. The body is doing it, anyway! **Your body is your temple!**

# Congratulations!

# You are the Living Intelligence in Action!

## 6. Self-Salvation is Our Life Elation!

**Attaining the Universal Intelligence is forming a bridge with the Divine.** This connection resonates in us through our **reading into meaningful coincidences** that we are learning to pay attention to. With your personal intelligence constantly changing, your consciousness enriching, and your will-power strengthening, ***you will start perceiving higher beings around you***, reasoningout very step with them, tuning your intuition up to their advice, and being able to finally view **every *meaningful coincidence*** as a sign of God, guiding your life.

***Higher Consciousness without any crowd-channeled controlis our ultimate destination in self- creation!***

I love the words by ***Alexander Blok***, a great Russia poet that I translated and keep quoting in my books. *A. Blok's* words characterize our insatiable desire to go beyond the terrestrial and do the impossible to become one with God.

**"I want to desperately live**

**To internalize what can be seen,**

**To celebrate the unforeseen,**

**To humanize the irreversible,**

**And realize the impossible!"**

*(Alexander Block)*

*"Inner peace can only come as a natural consequence of the Universal Enlightenment."(Dalai Lama).*

## With the Mission Accomplished, You Can DeclareYour Inner Freedom here and there!

# Good Luck on the Self-Installation Track!

# The Extra-Terrestrial Aims are Our Future Gains!

# Conclusion to the System Infusion

# Intelligence-
# Ascending is
# Self-Consciousness-
# Mending!

*Be a Thought-Evolved Being with the Universal Seeing!*

**Become Less Life-Negligent and More Life-Intelligent!**

## 1. Your Personal Pride is Inflated with the Digitally Intellectualized Might!

In sum, you realize, of course, that self-***installation is a lifelong process*** with two steps forward and one step back, or sometimes one step forward and two steps back. That is a natural process ofself-creation. The most important thing is ***not to get on the rocky road of self-corrode***. It is alwaysthere, but your self-consciousness is getting higher, stronger, and more refined. So, to be more life-fitting, keep repeating auto-suggestively:

# I am not life-negligent, I am life-intelligent!

When your spirituality is much higher, and your self-consciousness is more developed, you willfinally arrive at *the* ***Integral You,*** or "***a Star Man***" stage that John Baines predicts for us in his book "***The Stellar Man".*** The best of us will attain this stage in the unforeseeable future. Meanwhile, the change will be manifesting itself in you because you are following ***the objective path,*** not just the one, suggested by a certain Dr. Ray

***.I have verified this path all my professional life, and I can swear to you, it is right!***

Anyway, you no longer have judgment about anything and anyone. ***You are no longer like anyone!*** At this Super level, we all become one with life and the immortal spirit of Universal Intelligence which is God! ***The Universal intelligence is making your mind-body obey your willwhich is God's will! What an uplifting feeling that is!*** You will experience it at some point without fail! Your ardent desire to oversee yourself and attain full self-mastery will be instilled and constantly fortified, building up **your *inner fortress, your personal magnetic core*** where youwill find refuge any time you need help. Just the reminder of the formula of strength below will do the trick. ***Its inductive power is immense***!

**I Can Roam Any Terrain with God in My Vein!**

May this program become the conceptual blueprint of self-creation for the rest of your life. The basic thing for you is to do self-hypnosis consistently as a boost for the self-image and the nourishment for your willpower. Self-hypnosis helps synchronize thoughts and emotions and reeducate the process of thinking. Eventually, you will learn to think consequentially, not impulsively and compulsively because awareness or conscious assessment of self-reality will change your mental digestion for new frequencies of intelligence and self-consciousness in synch.With new, conscious responding to life, you will attract people that have the same gravitational force in their minds, and you'll become the ***Living Intelligence in action.***

# Only with God in Synch, you are What You Think!

## 2. The Holistic Education is the Solution for Our Evolution!

Our educational goal is to gradually move to the universal realm of extra-terrestrial communicationand become the citizens of the universal realm of the mind ***- the Source !***New amazing discoveriesare made every day thanks to the electronic intelligence in every area of science that is propellingus ***up-ward, on-ward, and God-ward***.

Apparently, the development of human consciousness is inseparable from the digital evolution thatis pushing us forward in terms of ***creating a new volume of our consciousness*** *that* will expand our communication with each other and the world around us inconceivably. *(See Electronic Intelligence below)*

## No doubt, Artificial intelligence is channeling our evolution!

New times need constructive, well-structured, properly channeled, emotionally empowered mindsthat will necessitate physical, emotional, mental, spiritual, and universal changes and propel us forward to the Star people level.

***The fantastic thought-machine of yours is not for the residence of moths!***

There is a great old Jewish saying that illustrates the idea of a conceptual mind-set that we need togenerate, ***"If there are only two courses of action, you should always take the third one."*** In ourcase, it's the self-development along the paths of the ***Holistic Education***, the education that provides the development of intelligence **in five levels** that will considerably expand our humanevolutionary outreach and raise our self-consciousness:

*Physical + Emotional + Mental + Spiritual + Universal!*

**Unconscious** → **Conscious** → **Superconscious**

*Self-Awareness – Self-Monitoring - Self-Installation – Self-Realization - Self-Salvation!*

Such education, backed up by an exponential growth of technology and supplemented by anirresistible urge for self-*installation in life, should follow* ***the holistic formula:***

## Synthesis – Analysis – Synthesis!

### Internalizing knowledge – Personalizing it – Externalizing it!

## Our Digitally Enhanced Human Essence is in the Holistic Renaissance!

# 3. Ten Stages of the Living Intelligence

The Universal Intelligence completes ***the Self-Actualization Pyramid*** of the essential levels of the Living Intelligence in the mental realm of the spiral of self-resurrection that I have outlinedfor you in most general terms above. All the stages are interconnected in the holistic system, presented stage by stage, *T*here are two more levels to complete after the mental one.

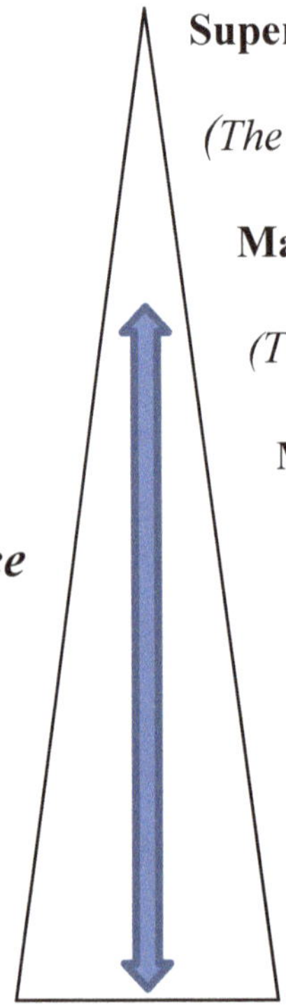

| | |
|---|---|
| ***10. Universal Intelligence*** | **Super - Level** of Consciousness **/ Step 10** |
| ***9. Spiritual Intelligence*** | *(The Universal Dimension )* **Step 9** |
| ***8. Social Intelligence*** | **Macro - Level / Step 8** |
| ***7. Cultural Intelligence*** | *(The Spiritual Dimension)* **Step 7** |
| ***6. Financial Intelligence*** | **Mezzo - Level / Step 6** |
| ***5. Professional /****Creative* ***/ Intelligence*** | *( The Mental Dimension* |
| ***4. Psychological Intelligence*** | **Meta - Level / Step 4** |
| ***3. Emotional Intelligence*** | *(The Emotional Dimension)* **Step 3** |
| ***2. Language Intelligence*** | **Mini - Level / Step 2** |
| ***1. General Intelligence*** *(Self-Genesis)* | *(The Physical Dimension)* **Step 1** |

**The Spiral of Self-Resurrection is completed!**

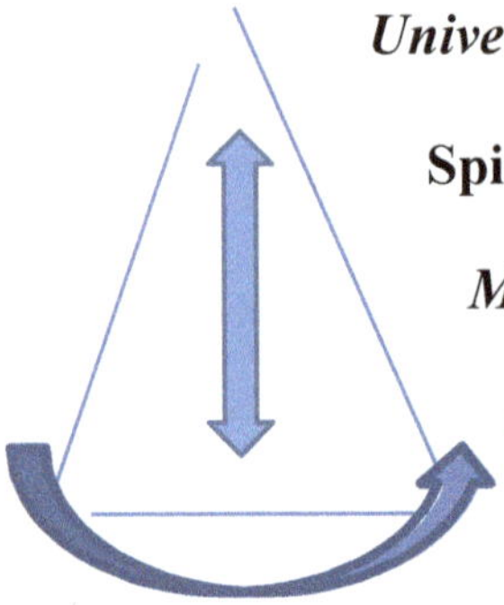

| | | |
|---|---|---|
| ***Universal level*** | **Super-Consciousness** | ***Self-Salvation*** |
| **Spiritual level** | **Self-Consciousness** | ***Self-Realization*** |
| ***Mental level*** | **Mind** | ***Self-Installation*** |
| ***Emotional level*** | **Spirit** | ***Self-Monitoring*** |
| ***Physical level*** | **Body** | ***Self-Awareness*** |

**Body+ Spirit+ Mind + Self-Consciousness+ Universal Consciousness =**

## A New, Holistic, Intellectually Spiritualized Cosmic Self!

## 4. Some Basic Tools for the Self-Refinary Rules

*1) Techlogical Renaissance + Science Renaissance + Self-Renaissace =*

**Your Technologically Enhanced Self-Acculturation!**

***2) The physical, emotional, mental, spiritual, and universal stratacomprise your life's Transformational Plato!***

***3) Intellectualize Your Emotions and Emotionalize the Mind!***

**Be One of a Kind!**

***4) There want' there, isn't, there won't ever be***

**Anyone like Me!**

**5) I KNOW WHO IAM, AND I KNOW WHO I AM NOT! I**

***I am a strong, calm, bold, and determined owner of my firm will!***

**I can…; I want to…; and I will…!**

***I am becoming better and better at reforming myself / etc. with each coming day!***

**6) Put Your Heart and Mind in Sync. Feel , but Think!**

**7) *"Will Your Life More!"*** *(Carl Yung)* ***That's the Law!***

**8) *To Succeed, Be Mind and Language-Fit!***

**9) To Be Inspired, Be Self-Inspiring!**

***10) Constantly ask yourself what the world could be and what you have done for it to become better.***

**Self-Renaissance is Your Chance!**

## 5. Oh, Gee; I'm the Quality Me!

*I am driving through the time and space with intelligence, awareness,love, success and health on my interface!*

*Be sure to declare:*

# I am Physically Beautiful,

# Emotionally Invincible,

# Mentally Unbeatable

# And Spiritually – Free!

# I Am a New, Conscious Me!

***Self-Synthesis*** → ***Self-Analysis*** → ***Self-Synthesis*****!**

**Long Live the Belief in Myself without If!**

**Carpe Diem! Some Day is Today!**

# Post Thought

## "I Have a Dream!" (*Martin Luther King Jr.)*

When we arrived in the USA, 28 years ago, there were many things that we lacked in the former Soviet Union of those destructive times, but the country of plenty lacked one thing that my 13-year old daughter noticed very soon and told me one day after she came home from Mamaroneck school. in Westchester. -*"**Mom, I know my dream now. I will start my own business in this country. I will build the center for kids! They have no Pioneer Cantersfor kids, where I have learnt folk-dancing, ballet - dancing, sewing, fishing, swimming, space-flying, and toy-making. Poor kids. They don't have such free centers for kids here. Nor do they have anything for teenagers. What should I do here? Where do I go after school? My school peers talk only about parties, bars, sex, and sleepovers."***

I was glad that she started thinking in the American way business-wise, but I was truly upsetthat she was right about the second part of her observation that very soon became the reality and a nightmare for me. The life changed her dream, and even though she became a computer-designer and got seven books for kids published, she did not manage to bridge her dream and the reality. She had never spoken about it again. But I have never forgotten thatconversation and ***the spark in Yolanta's*** eyes because I saw the same spark in the eyes of many of my students. I inspire them to never betray their dreams and become what they aspireto be, not just what secures their life As a matter of fact, ***my five books on self-awareness, self- monitoring, self-installation, self-realization, and self-salvation*** in life, are inspired by my daughter's unrealized dream and my students desire to accomplish theirs.

## "LIFE IS A SPIRAL, NOT A CIRCLE!"

My dream is to have a free developmental ***center "Jupiter"*** that will be structured in five rings - ***physical, emotional, mental, spiritual, and universal***, presenting five ***space stations:***

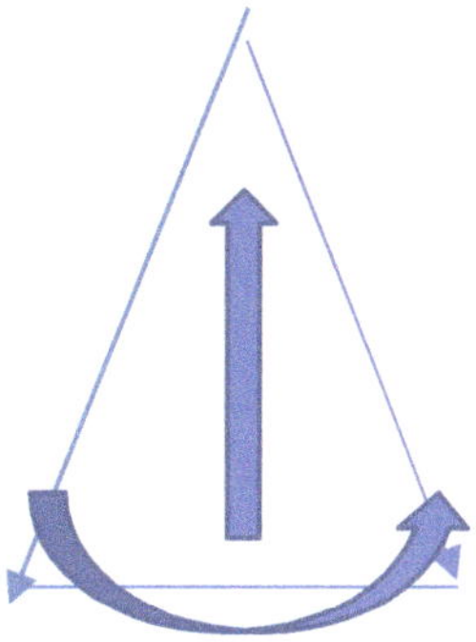

**5. Universal Space Station**

**4. Spiritual Space Station**

**3. Mental Space Station**

**2. Emotional Space Station**

**1. Physical Space Station**

Wouldn't it be great to have the Physical Space Station for the kids and teenager to develop their physical abilities, like it was in ***the ancient Greek school for Spartans***. They will enrichtheir ***cognitive skills*** and learn many new interesting things of their choice, as well as train their digital ***skills*** in the most creative ways. *(The book "I am Free to Be the Best of Me!")*

Next, at **the Emotional Space Station**, the kids will learn about an array of beautiful humanemotions and how to handle their Amygdala glad with the help of the best movies, tales, works of art, and beautiful classic music There will be an instructor- an ***emotionalist*** *and* ***robot -friends*** to back them up at the moment of dismay. *(the book "Soul-Refining")*

The holistic pyramid of essential ***vistas of intelligence*** will be at their fingerprints when theychose to spiral up **to the Mental Station** to fill up the gaps in the levels of intelligence that this book features. ***They will enhance their awareness by connecting their two brains*** and boosting their creativity and ascertaining their dream of what they want to become *(the book "Living Intelligence or the Art of Becoming!")*

The most inspiring knowledge of the human evolutionary ***spiritual growth*** will await them ***atthe*** **Spiritual Space Station,** without any preaching, but insightful, interesting teaching to shape their S***piritualized Intelligence*** that is operating the entire space. *( the book " Self- Taming!")*

The space-time journey into the impossible will be completed ***at*** **the Universal Station,** witha breath-taking flight of imagination in the ***top observatory*** + ***a virtual space station*** that willtake the kids on interstellar travels into space. *( the book " Beyond the Terrestrial!)*

**And the best thing of all** - there will be no obligation, no words: "***You are supposed to..."/"You need to pay first!" That's not your station!" You need to first pay this..."*** The possibilities are limitless, as well as our **INCREDIBLE NOW** that is, in fact, our wonderousfuture if we allow the dreams of our kids come true!

***The choices we make dictate the life we live!***

**"If you want to shine like the Sun, first burn like the Sun."**

*(Abdul Kalam)*

# And That, in Every Air Sip, is Our Universal Trip!

## Dr. Ray with Her Inspirational Say!

*1. "**Emotional Diplomacy** or **Follow the Bliss of the Uncatchable Is!"/ Editorial** LEIRIS, NewYork, USA,2010*

*2. **"Four Dimensions of a Soul**" (Auto-Suggestive Psychology in Russian) / LEIRIS Publishing,New York, USA,2011*

*3. **"Americanize Your Language, Emotionalize Your Speech!"** / Nova Press, USA, 2011*

*4. **"It Too Shall Pass!"** (Inspirational Psychology for Self-Ecology) / Xlibris, 2012*

*5. **"I am Strong in My Spirit!"** (Inspirational Boosters in Russian) / Xlibris, 2015.*

*6. **"Language Intelligence or Universal English"** (Method of the Right Language Behavior), **Book One** /Xlibris, 2013 - Also, Stonewallpress,2019*

*7. **"Language Intelligence or Universal English**" **(**Remedy Your Language Habits," **Book Two** /Xlibris, 2013 – Also, Stonewallpress,2019*

*8. **"Language Intelligence or Universal English,"** (Remedy Your Speech Skills) **Book Three** /Xlibris, 2013- Also, Stonewallpress,2019*

*9. **My Solar System** (Inspirational Psychology for Self-Ecology), Second Edition- Page Turner,Press and Media, 2019*

*10. **Language Intelligence** (three books in one) Stonewall Press, 2019*

### Books on the Holistic Self-Resurrection:

*11. **"I Am Free to Be the Best of Me!"- (Physical Dimension) -** Toplinkpublishing.com. Sept.2017) – Second Edition, Book Whip, 2019*

*12. **Soul-Refining! (Emotional Dimension) (**Toplinkpublishing.com. May 2017) - Second Edition2ook Whip, 2019*

*13. **"Living Intelligence or the Art of Becoming**" (Mental **Dimension)-** (A New Paradigm of Self-Creation) Xlibris, 2015 - Second Edition, Bookwhip,2019*

*14. **"Self-Taming" (**Life-Gaining is in Self-Taming**! / Spiritual Dimension)-** Book Whip, 2019*

*15. **" Beyond the Terrestrial!"** (Be the Station for Self-Inspiration**!) - (Universal Dimension)** Xlibris, June 2016- Second Edition -, Book Whip, 2019 / - URLink Print and Media, 2019*

*16. **The State of Love from the Above** (**Love Ecology for Self-Psychology**) Book Whip, 2019*

*17. "Love-Ecology"*

*18. "Self-Worth"- Parchment, new Your,2020*

*19.* **"Self-Renaissance" / Workbook press, Las Vegus, 2020**

**20" Soul-Symmetry!"/ *Clever Publication, Canada, 2021***

**21. "Dis-Entanglement!"/ Ivy Lit Press, New York, 2022**

*www. Language – fitness.com / email; - dr. rimaletta@gmail.com*

*Pictures and designs are by **Yolanta Lensky** –*

***(203) 212-2673***

***"There are Two Most Important Days in your Life –***

***the day you were born, and the day you realized What For."***

*(Mark Twain)*

**Don't Life-Sway,**

**Somewhere and Someday are Here, today!**

# Life is Going on, and it is great to Have Been Born!

www.ingramcontent.com/pod-product-compliance
Ingram Content Group UK Ltd.
Pitfield, Milton Keynes, MK11 3LW, UK
UKHW061350110926
15112UKWH00028B/223

9 781960 159793